Empowering Educational Leaders

Sabiha Dulay (ed.)

Empowering Educational Leaders

How to thrive in a volatile, uncertain, complex and ambiguous world

PETER LANG

Lausanne - Berlin - Bruxelles - Chennai - New York – Oxford

Library of Congress Cataloging-in-Publication
This book has been requested for registration in the Library of Congress
CIP catalog of the Library of Congress.

Bibliographic Information published by the Deutsche Nationalbibliothek
The Deutsche Nationalbibliothek lists this publication in the Deutsche
Nationalbibliografie; detailed bibliographic data is available in the internet
at http://dnb.d-nb.de.

ISBN 978-3-631-89596-2 (Print)
E-ISBN 978-3-631-90673-6 (E-PDF)
E-ISBN 978-3-631-90674-3 (E-PUB)
10.3726/b21086

Published by Peter Lang GmbH, Berlin, Deutschland

info@peterlang.com - www.peterlang.com

Editor's Preface

We live in a rapidly changing world with its unique challenges and opportunities for growth and innovation. The world is changing and transforming institutions and leaders. Thus, it has become increasingly challenging for people to keep up with the technological, economic, and social advances in this rapidly changing world. This world is defined as VUCA which stands for volatility, uncertainty, complexity, and ambiguity. The term VUCA describes the changing nature of business environment and has gained importance in recent years. To be able to adjust changes, complexities, and deal with them effectively, leaders need to be prepared for the crises and opportunities that await them. Thus, this book aims to develop educational leaders for volatile, uncertain, complex, and ambiguous world, and help them gain a new mindset to manage their educational organizations. Many of us agree that we need different tools, a new set of skills and attempts to deal with future challenges and opportunities. It is obvious that traditional ways of leadership and management are not working anymore. Leaders need to understand the context and implement effective leadership practices with a new mindset. That's why this book provides current information on the new leadership skills and enhancing VUCA readiness of educational leaders. This book can be seen as a practical guide for educational leaders by discussing the insights and perspectives that would help them to lead schools. Whereby, the educational leaders understand and apply these guiding principles, they will adapt to changes quickly and be ready to follow future trends as an effective leader in the VUCA world. In this context, our book is thought to be important for practitioners, leaders and teachers who need new skills and guidance on how to manage with an effective, agile, and authentic manner. In the light of the current discussions, this book tries to relate theory and practice through discussing leadership in a changing world with practical suggestions.

Acknowledgments

I would like to thank all the authors for their time, energy, interest and contribution to get the book completed on time, which would not have been possible without them. I wish to thank Prof. Dr. Şirin Karadeniz, the President of Bahçeşehir University for her academic leadership, encouragement, the supportive environment that she provided for us. Lastly, I would also like to thank the publishers, Peter Lang for their support and agreeing to publish this book.

Sabiha Dulay, PhD Bahçeşehir University,
Istanbul, Turkey
May 2023

Contents

Introduction

This book covers four parts and 11 chapters as follows:

Part 1, including one chapter, focuses on the complexity of VUCA world, describes the challenges and opportunities for leaders and offers clarity for leaders to survive in a VUCA world.

Part 2, including four chapters, offers new skills to help leaders thrive in the VUCA world through leading change. Chap. 2, aims to examine the role of authentic leadership which serves as the foundation for various other positive leadership styles in the current VUCA environment at schools. This chapter has underlined the facilitating role of authentic leaders to inspire people to adopt new viewpoints, question established ideas, and cultivate flexibility, inventiveness, and self-assurance. In today's world, one of the competencies that leaders need to develop is to manage the current situation and to be ready to act agile against the change in the face of uncertainty. Thus, Chap. 3 aims to explain the concept of agile leadership as an important leadership model in the VUCA world. Chap. 4 is dedicated to explaining the fundamentals of quantum leadership and characteristics of quantum leaders who inspire others to thrive during periods of great change and uncertainty by drawing upon the theory of chaos and complexity. In Chap. 5, the authors provide insights on promoting leaders' resilience by identifying the lessons learned from the COVID-19 crisis as a sample model.

Part 3, including three chapters, explores the ways for building communication and collaboration. Under this heading, Chap. 6 examines the complex and diverse VUCA environment and discusses the necessary skills and the mindset that leaders should have to manage diversity and build a collaborative school culture for equity and inclusion. Chap. 7 covers the five disciplines, 11 Laws and truths of the fifth discipline/system thinking that can be used as a guideline helping educational leaders create a more hopeful future for both teachers and children. Risk management is one of the most important processes affecting enterprises that leads them to success and sustainability. Thus, Chap. 8 begins with a definition of risk, and goes on with risk management process and strategies.

Part 4, consisting of three chapters, identifies the digital transformation in which schools are. Chap. 9 focuses on developing new-age educational leaders and leadership with respect to Industry 5.0 in order to help facilitate these changes in educational environments. Therefore, history of industrial revolutions, reflections on daily life, new-age leadership and digital transformation, and emergence of new professions and skills are the main topic of interest in this chapter. Chap. 10 focuses on three main issues. First, the skill gap phenomenon between universities and the business world is examined. Second, leaders' professional development and digital workplace learning challenge are discussed and last, Massive Open Online Courses (MOOCs) as lifelong learning supporters are clarified. The fact that educational leaders who dominate artificial intelligence, cloud technologies, big data and other trending topics show visionary leadership and create a digital-age learning culture in their environments. Chap. 11 aims to help educational leaders understand how to use and integrate artificial intelligence, big data and cloud technologies within their organizations and how they will create changes when used. In conclusion, each chapter concludes with recommendations and strategies for educational leaders in complex and constantly changing world, called VUCA.

Part 1: VUCA World

Dr. Sabiha Dulay

Let the Fog Cloud Dissipate: Offering Clarity to Educational Leaders in a VUCA World

Abstract

This chapter aims to provide a framework for today's constantly changing environment known as VUCA. The acronym VUCA which stands for volatility, uncertainty, complexity, and ambiguity, reflects an increasingly unstable and rapidly changing business world (Lawrence, 2013, p. 2). This unstable nature of the world has been transforming schools and educational leaders. Therefore, not surprisingly, trends and developments affect educational organizations and create a considerable need for qualified educational leaders. VUCA requires new sets of skills, practices, and attempts to overcome problems, thus, adaptation to VUCA world might be difficult for leaders with traditional leadership skills. First, leaders need to understand the challenges of the VUCA world and gain a new mindset to manage the organizations, and to navigate the complexities. That's why this chapter provides current information on highlighting the complexity and challenges of the VUCA environment, providing a deeper understanding about the new leadership skills, and enhancing VUCA readiness of educational leaders. We need this discussion to underline the importance of VUCA readiness of educational leaders since they need new skills, mindsets, and methods to drive growth, stay ahead of the curve, and lead their organizations to succeed. Within this main goal, which is to develop educational leaders for a volatile, uncertain, complex, and ambiguous world, four themes (Understanding VUCA world, Challenges and Opportunities for Leaders in a VUCA World, How can Leaders Survive and Lead in a VUCA World, Learning Needs and Skills of Educational Leaders in a VUCA World) are discussed specifically with a holistic view including practical suggestions for educational leaders.

Keywords: VUCA, VUCA-readiness, leadership competencies, educational leaders

> *"The greatest danger in times of turbulence is not the turbulence; it is to act with yesterday's logic."*
>
> *—Drucker (1980)*

Educational organizations in today's world are now facing inevitable changes, which call for effective leaders to survive and succeed. In this evolving environment, leadership has taken on an entirely new angle. Now, leaders should not only manage schools, but also have leadership skills that enable them to be flexible, innovative, and adaptive to changing conditions. This world of rapid changes with new perspectives is known as the VUCA world. The acronym VUCA which stands for volatility, uncertainty, complexity, and ambiguity, reflects an increasingly unstable and rapidly changing business world (Lawrence, 2013, p. 2). The term was originally coined by the U.S. military in the 1990s to describe the challenges and the changing environment caused by the post-Cold War world (Bennett & Lemoine, 2014). It was first used in 1985 by Bennis and Nanus to describe the new situation that arose in the 1990s because of the fall of the USSR. Then, it was quickly introduced to different areas to understand how this unstable environment force organization to change. It can easily be seen in the story of Kodak and Polaroid or Apple, Google, 3M and Amazon. Kodak and Polaroid refused to adapt to new normal triggered by the digital transformation and failed, while the successful businesses like Apple, Google, 3M, and Amazon that seem to be able to quickly adapt to developments and integrate technology into their environments. The significance of leading and adapting to the changing environment were once again highlighted by these outstanding cases. On this basis, we aim to introduce readers to understand VUCA world, what challenges and opportunities they might face, and what kind of leadership skills they need to succeed.

Understanding VUCA World

What do we know about VUCA? Why do we need to understand and be prepared for it? Clearly, answers to these questions are far from simple. Moreover, the answers are rather complex because the questions are complex. To understand what leaders need to be effective in VUCA, we must first understand what is meant by the acronym VUCA. Although the term first appeared in the military, it was quickly transferred into different areas including business, leadership and management. In connected with this basis, basically, it can be easily said that there were many periods as an example of VUCA, including global climate debates (Johansen & Euchner, 2013), digitalization (Çetin & Karsantık, 2022; Lawrence, 2013), financial turbulences (Sullivan, 2012), and COVID-19 outbreak (Dhillon & Nguyen, 2020). The chaotic and rapidly changing environment in which we live has called for new normal (Lawrence, 2013), which possesses a distinctive feature such as being more interconnected, unstable, severe economic challenges, and unequal income distribution (Heinonen et al., 2017), thus, the quantity of current debates on the need for qualified leaders is escalating with each passing day.

VUCA not only brings out the constant changes in workplaces (Mahlomotja, 2018; Mayer & Wilke, 2022), but also offers opportunities and insights into challenges of rapidly changing and unpredictable world which have been constructed by globalization and technology (Millar et al., 2018). That's why

VUCA has become more prevalent in leadership, management, and organizational studies (Kok & van den Heuvel, 2019; Mahlomotja, 2018). Otherwise, if the leaders are not ready for the VUCA environment, the challenge may negatively impact their well-being and mental health (Millar et al., 2018). Therefore, leaders must understand VUCA environment, its components, and be prepared for the opportunities, challenges, and crises caused by a VUCA world. But first, it is vital for leaders to define and discuss the basic features of VUCA. It is an acronym standing for Volatility, Uncertainty, Complexity, and Ambiguity. Each component has different characteristics and requirements which are briefly explained here:

(i) *Volatility:* The "V" in the VUCA is for volatile, referring to the nature, speed, volume, magnitude and dynamics of change (Horney et al., 2010). In a volatile environment, things can change rapidly and unpredictably, making it difficult to plan and anticipate the future. Accordingly, it demands leaders respond quickly to continuous changes that are unpredictable and beyond their control.

(ii) *Uncertainty:* The "U" in the VUCA is for uncertainty, referring to a lack of predictability and the impossibility to know with absolute certainty what will occur in the future. It might be challenging for leaders to decide which of several potential outcomes in an unclear situation is likely to occur.

(iii) *Complexity:* The "C" in the VUCA is for complexity, referring to the interconnectivity of systems and situations. In a complex environment, there are often many different factors and variables at play, and it can be difficult to understand how they all interact. Thus, one must put great effort to collect, and understand all the relevant information (Ramakrishnan, 2021).

(iv) *Ambiguity:* The "A" in the VUCA is for ambiguity, referring to a lack of clarity or understanding. In an ambiguous environment, it can be difficult to determine what is happening, what it means, or what actions to take. As the causal links are unclear, the real issues within threats and opportunities are not clearly seized (Ramakrishnan, 2021), decision-making has become significantly more challenging (Sinha & Sinha, 2020) in today's ambiguous world.

Each world of the acronym VUCA represents a different type of challenges that leaders should know and that make their jobs more difficult than before (Bennett & Lemoine, 2014). Therefore, leaders should be aware that each component requires different response to its challenges (Thebe, 2017). Volatility, which means that leaders must be able to respond quickly to unexpected changes, causes instability for organizations. Uncertainty, which refers to leaders lack a clear understanding and knowledge of what the future holds, creates problems in decision-making. Due to the interconnectedness of the factors within complexity, leaders must be able to navigate a wide range of variables at the same time. Lastly, leaders must be able to deal with situations that are unclear or difficult to interpret in an ambiguous world. To manage these challenges, and navigate conflicts, leadership in a VUCA world requires new skills. They need to be ready for the VUCA environment and its challenges (Rimita, 2019) which requires agility, flexibility, and the ability to innovate and think creatively. Not only the leaders, but also organizations must be able to respond appropriately to changing conditions, prepare for potential crises, and successfully navigate uncertain settings in a VUCA world.

As Kok and van den Heuvel (2019) emphasized that we are witnessing the greatest rapid change in human history, posing various leadership challenges. These challenges can be daunting for leaders and organizations, but they also present opportunities for growth and innovation. In this regard, we'll first go through the challenges and opportunities of VUCA, before talking about the strategies that will help them succeed in the changing environment.

Challenges and Opportunities for Leaders in a VUCA World

Leaders are now facing uniquely difficult and complex scenarios (Coopersmith, 2022) and challenges in today's swiftly changing working environment in which they must be able to adapt and respond accordingly. The first leadership challenge is basically the VUCA itself, that is, the rapid, complex and unpredictable conditions. The unpredictable nature of VUCA World causes leaders to have more stress and various mental health problems (Millar et al., 2018). Meanwhile, this unpredictability also poses a paradigm shift in leadership and management which leaders must be able to aware of its consequences. To better understand this paradigm shift, it will be useful to comprehend the changing circumstances discussed by Mack and Jungen (2016). Based on their findings, trying harder does not work anymore since the challenges of today and tomorrow cannot be understood with yesterday's solutions. In this new environment, it is not possible to define the right perspective or stay the same instead we need different perspectives and to be adaptive to understand the big picture. Knowledge and collaboration will be the key for anyone to make innovations possible (Mack & Jungen, 2016). Considering the changing environment defined by Mack and Jungen (2016), it is as clear as day that there is a rising need to prepare leaders for VUCA environments, which poses another challenge. This is how to be an effective leader who can survive in unpredictable and rapidly changing conditions which leaders need to make different adjustments for the sustainability of the organization (Pertiwi et al., 2022).

VUCA is also regarded as a challenging environment within the educational sphere (Suzana & Mohd Amir Shaukhi, 2020), and leaders thus face many of the same challenges as business leaders in a VUCA world (Bennett & Lemoine, 2014). For example, the year 2020 can thus be seen as an example of how the VUCA era has evolved (Dhillon & Nguyen, 2020), which changed the world of education forever (Li & Lalani, 2020), and more is required than traditional leadership to survive (Cinnioğlu, 2020; Johansen, 2012; Kukreja, 2019) in this constantly changing and complex world. The COVID-19 pandemic has introduced significant challenges for educators and students, requiring them to be agile and adaptive in the face of unexpected and rapidly changing circumstances (Tadesse & Muluye, 2020). Leaders should respond to the demands of the educational environment with different strategies. According to Panthalookaran (2022), leaders could overcome volatility with vision and autonomous learning based on exploration and experiments, uncertainty with understanding and lifelong learning emphasizing the process of learning rather than knowledge product, complexity with courage and venturesome learning building up courage and confidence to implement their ideas, ambiguity with adaptability and flexible learning characterized by a learning-unlearning—relearning triad, as given in Fig. 1.

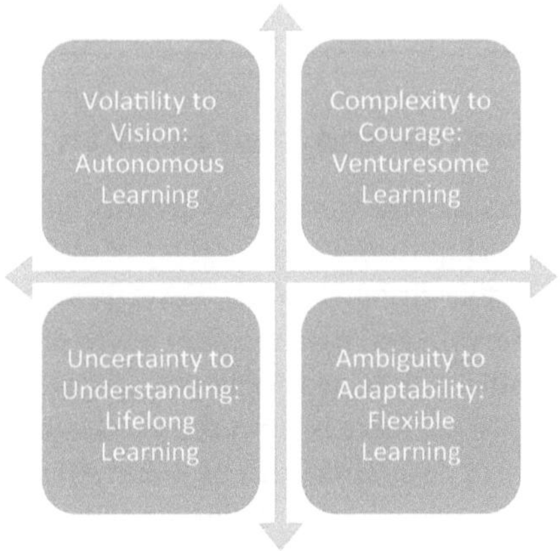

Fig. 1. Features of Education in a VUCA-Driven World (Panthalookaran, 2022)

The other paradigm shaping education is Education 4.0, which mainly integrates technology into teaching and learning process. This changing nature of the educational environment with the COVID-19 pandemic, Education 4.0, developments and trends in teaching and learning reveals a need for effective and innovative leaders to stay ahead of the curve. They thus need to gain new skills or new mindsets to successfully lead in today's realities (Petrie, 2011). However, White (2022) addresses the issue that educational leaders are unprepared for VUCA environments due to the lack of knowledge of new educational leadership skills. They are ill-prepared and use traditional leadership tools which prevent them from overcoming the obstacles (Rimita, 2019; Rimita et al., 2020; Thebe, 2017). This is another of the biggest challenges for today's leaders face.

The main question needs to be answered here is not what challenges leaders face, but how to turn these challenges into opportunities. The changes created by the VUCA world force leaders to transform and offer the opportunity to acquire new skills. Leaders can learn how to adapt and respond to changes immediately to progress (Mayer & Wilke, 2022). They can encourage stakeholders to introduce new perspectives, think creatively with more powerful ideas, maintain tolerance, foster innovation and confidence (Chawla & Lenka, 2018). Additionally, leaders can gain the opportunity to develop a clear vision and deep understanding of the environment in which they live and to maintain agility and adaptability in time of rapid change. Therefore, understanding VUCA challenges and developing leaders for this environment may equip leaders with technical, structural, and psychological tools they need (Rimita et al., 2020) to lead successfully. So, they need to understand the VUCA environment, enhance their VUCA readiness to solve problems and empower themselves to embrace future challenges and opportunities (Suzana & Sazali, 2013) to learn how to survive.

How Can Leaders Survive and Lead in a VUCA World?

Scholars and practitioners are still working to identify effective leadership (Avolio, 2011) and how to develop future leaders (Day et al., 2014; Delves, 2022) for uncertain and unpredictable circumstances. Nevertheless, it is obvious that the traditional forms of leadership and management based on control and hierarchy are not working to handle this increasingly volatile, complex and ambiguous world anymore. As the pivotal force behind the successful organization, leaders need to think out of the box to adapt to this new and changing environment and provide a viable direction to go through the messes (Ramakrishnan, 2021).

The development of new mindsets, creative thinking, and transformed leadership roles are vital to dance with reality and to embrace ambiguity. Hence, the focus should be the transformational change in leadership role and the new mindset they need to be successful (Delves, 2022). Similarly, Dhillon and Nuyen (2020) call attention to need for leaders with the necessary skills to support this transformation instead of being an obstacle in a VUCA context. Thus, leaders are not allowed to be passive in this changing world (Johansen & Euchner, 2013), they should abandon old leadership practices (Lang & Sutton, 2022), and develop a strategic approach that will enable their organization to succeed (Dhillon & Nguyen, 2020) in the long run. The new leaders of today's evolving world are required to be more participative (Dhiman, 2022), more agile and adaptive (Pertiwi et al., 2022; Yurdasever & Fidan, 2020), more authentic and emotionally intelligent (Delves, 2022; Mahlomotja, 2018), and more responsible (Sarkar, 2016). Similarly, new leadership development models have been an increased demand in educational environment since managing todays' school is getting more complicated as a result of the VUCA challenges.

To manage challenges and opportunities effectively, educational leaders must adopt a proactive and adaptive approach to leadership that prioritizes innovation, collaboration, and engagement (Hargreaves & Fink, 2006). By adopting an adaptive and proactive approach to leadership, they can position their organizations for long-term success in an unpredictable and complex environment. In order to be effective leaders in today's educational environment, leaders could also adapt the strategy called VUCA Prime by Johansen (2009) to learning environments. According to his framework, leaders should face volatility with vision, overcome uncertainty with understanding, solve complexity with clarity, and fight ambiguity with agility.

(i) *Face Volatility with Vision:* Leaders with clear vision of where they want their organization to be better fight with volatile environmental changes (Lawrence, 2013). It is thus important for educational leaders to develop a clear and shared vision and convey it to all stakeholders.
(ii) *Overcome Uncertainty with Understanding* Leaders need to listen to others to develop new ways of understanding. Defining what are the knowable and unknowable facts is a good start in helping people to develop clarity around (Sharma, 2019). Educational leaders should keep both formal and informal communication channels open and obtain information from inside and outside networks of schools.
(iii) *Solve Complexity with Clarity:* Clear communication can enable leaders to understand what happened around during crisis or complex situations. It is important for leaders to communicate and collaborate by clarifying what is knowable and unknowable since complex environments are too

complicated to be understood by a single leader. Building trust-based relationships with teachers, students, parents, and other stakeholders in schools can help educational leaders to foster collaboration and navigate complexity with clarity.

(iv) *Fight Ambiguity with Agility:* Educational leaders should encourage flexibility, adaptability, agility and make plans but be prepared to change them. If the situation is not clear, then leaders need to have the flexibility to respond to whatever occurs (Sharma, 2019).

Learning Needs of Educational Leaders in a VUCA World

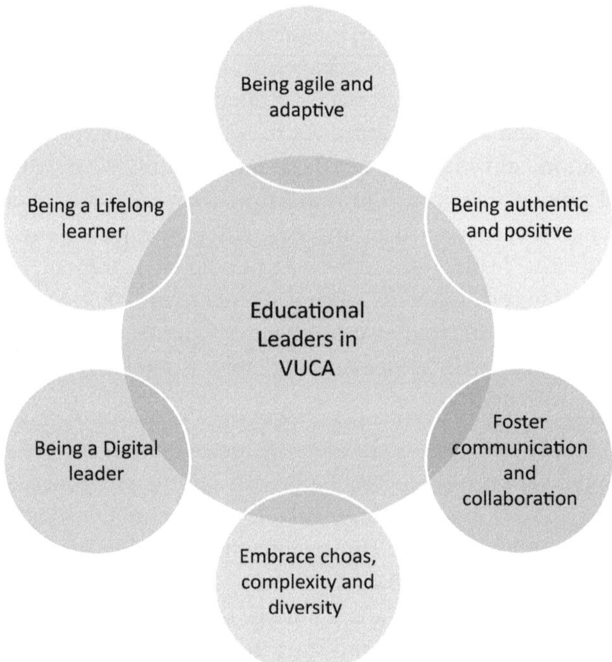

Fig. 2. Competencies of Educational Leaders in a VUCA-Driven World

As we discussed before, the new leader and perspective or paradigm shift on management are essential for today's complex educational environment. The new paradigm shift on leadership and management should be based on the ability to manage constant change, foster collaboration and be aware of technological advances rather than stability and hierarchies (Carnal, 2003). Time has changed but leaders are not ready for this changing and complex environment. It will thus be important to determine the necessary leadership competencies first. We, in this book, try to provide a framework to develop and prepare educational leaders for the challenges and opportunities of today's changing environment. As given in Fig. 2, we are discussing and suggesting the competencies that educational leaders need to have to lead their schools for growth and innovation.

Today's educational leaders need to develop themselves to be effective leaders of future schools. They must be able to embrace a growth mindset which helps them to see challenges as opportunities for growth and learning rather than an obstacle to overcome. With this mindset, they can encourage their teams to take risks and learn from failure (Dweck, 2008). After adopting a growth mindset, they need to acquire different competencies as follows:

(i) *Being Agile and Adaptive:* Agility and adaptation are key components of effective leadership in highly complex contexts. Educational leaders can successfully fight ambiguity with agility by responding quickly to changing circumstances (Lawrence, 2013).

(ii) *Being Authentic and Positive:* Educational leaders need to be authentic and positive to encourage and empower followers and foster innovation and flexibility. They as authentic leaders can be aware of the environment in which they work, have high moral standards, positive and optimistic.

(iii) *Foster Communication and Collaboration:* Communication and collaboration are the fundamental pillars of educational administration in the new paradigm. VUCA environment requires the integration of various skills, knowledge bases and capacities both within and without the organization and this could be a challenge if one was not collaborative as a leader (Mahlomotja, 2018). According to Kin and Karem's study (2019), educational leaders believe that improving schools should be more of a group effort than an individual one today. To be an effective educational leader during a time of crisis, they must possess effective communication skills (Kaivo-oja & Lauraeus, 2018; Mayer & Wilke, 2022; Smith & Riley, 2012), prioritize building relationships with students, teachers, parents, and other stakeholders (Hargreaves & Fink, 2006), and engage everyone in dealing with challenges and crises (Sharma, 2019) of VUCA.

(iv) *Embrace Chaos, Complexity and Diversity:* Educational leaders should embrace and know how to navigate chaos, complexity, and diversity, which makes everything simpler and clear. The simple and best way to respond to uncertainty and complexity is to gather information and to make sure that everyone has the same level of understanding. It is thus important to communicate and collaborate by clarifying what is knowable and unknowable. The school has been a more diverse place day by day; thus, leaders should know how to manage diversity and construct a safe and welcoming environment for everyone to live together. Educational leaders have a significant role in identifying the problems that can exist in diverse school environments, raising awareness of the similarities and differences between various cultures, encouraging and helping students in accepting others with tolerance and respect.

(v) *Being a Digital Leader:* Lastly but most importantly, educational leaders should display digital leadership by using technology in both administrative and instructional works (AlAjmi, 2022). In this regard, educational leaders should stay up to date with emerging technologies and use them to improve teaching and learning (Fullan, 2021), since these powerful tools enable leaders to connect with individuals in the schools, gather and manage big data, and increase innovation. By following trends and development in technology, educational leaders can understand how to use and integrate artificial intelligence, big data and cloud technologies within their organizations in a way that allows both students and teachers to progress in a supportive learning environment.

(vi) *Being a Lifelong Learner:* All these competencies call for an educational leader who is a lifelong learner who prioritizes professional and personal development. Leaders should be lifelong learners and prioritize both personal and professional development in the VUCA world requiring continuous adaptation, innovation, and learning to stay ahead of the curve. Lifelong learning is essential for leaders in a VUCA World, if not, they can miss the opportunities for innovation and growth. Many well-known and successful leaders from different sectors including education take online courses and also encourage their followers to follow the trends for personal and professional development.

Conclusion

This chapter aims to contribute to the knowledge base in VUCA world and underlines the importance of VUCA readiness of educational leaders since they need new skills, mindsets, and methods to drive growth, stay ahead of the curve, and lead their organizations to succeed. That's why this chapter discusses the challenges and opportunities that leaders face in today's constantly challenging environment known as VUCA first. VUCA not only brings out the constant changes in workplaces (Mahlomotja, 2018; Mayer & Wilke, 2022), but also offers opportunities and insights into challenges of rapidly changing and unpredictable world (Millar et al., 2018). On this basis, it is vital for leaders to understand VUCA with challenges and opportunities and learn how to adapt to this new environment for growth and innovation. Based on the studies and their findings, it can be inferred that leaders cannot be effective with traditional leadership skills and approaches. In fact, they are not ready for this dynamically changing working and social environments since they lack the necessary skills and knowledge for effective VUCA leadership. So that, they might encounter various obstacles to handle with VUCA and fail to follow the trends, developments, and opportunities for growth. If they are not ready for the VUCA environment, they more likely experience stress and mental health as Millar et al. (2018) underline. Especially when we consider the examples related to VUCA, we have experienced recently such as financial turbulences, climate change, digitalization and COVID-19 pandemic, it has become more important than ever to prepare leaders for a changing and increasingly complex world.

The main question answered in this chapter was not how to manage these challenges, but how to turn them into opportunities. Leaders can gain the opportunity to develop a clear vision and deep understanding of the environment in which they live and to maintain agility and adaptability in time of rapid change. Like the successful businesses such as Apple, Google, 3M, and Amazon did, they must be able to quickly adapt to technological developments and new trends in school management in order to be effective leaders of future schools. For this reason, they need to embrace a growth mindset which helps them to see challenges as opportunities for growth and learning rather than an obstacle to overcome. With this mindset, they can encourage their teams to take risks and learn from failure (Dweck, 2008). Consequently, this chapter ends with the following suggestions on the competencies that leaders need in a VUCA word:

- Educational leaders need to be agile and adaptive to fight with ambiguity and respond quickly to changing circumstances.
- Educational leaders need to be authentic and positive to encourage and empower followers and foster innovation and flexibility. They as authentic leaders can be aware of the environment in which they work, have high moral standards, be positive and optimistic.
- Educational leaders should foster communication and collaboration as the fundamental pillars of educational administration in the new paradigm. To be an effective educational leader during a time of crisis, they must possess

effective communication skills (Kaivo-oja & Lauraeus, 2018; Mayer & Wilke, 2022; Smith & Riley, 2012), and prioritize building relationships with students, teachers, parents, and other stakeholders (Hargreaves & Fink, 2006).

- Educational leaders should embrace and know how to navigate chaos, complexity, and diversity, which makes everything simpler and clear. The school has been a more diverse place day by day; thus, leaders should know how to manage diversity and construct a safe and welcoming environment for everyone to live together.
- By following trends and development in technology, educational leaders can understand how to use and integrate artificial intelligence, big data and cloud technologies within their organizations in a way that allows both students and teachers to progress in a supportive learning environment.
- Leaders should be lifelong learners and prioritize both personal and professional development in the VUCA world which requires continuous adaptation, innovation, and learning to stay ahead of the curve.

References

AlAjmi, M. K. (2022). The impact of digital leadership on teachers' technology integration during the COVID-19 pandemic in Kuwait. *International Journal of Educational Research, 112*, 101928. https://doi.org/10.1016/j.ijer.2022.101928

Avolio, B. J. (2011). *Full range leadership development*. Sage.

Bennett, N., & Lemoine, G. J. (2014a). What VUCA really means for you. *Harvard Business Review, 92*(1/2), 27. https://ssrn.com/abstract=2389563

Bennett, N., & Lemoine, G. J. (2014b). What a difference a word makes: Understanding threats to performance in a VUCA world. *Business Horizons, 57*(3), 311–317. https://doi.org/10.1016/j.bushor.2014.01.001

Carnall, C. (2003). *Managing change in organizations*. Pearson Education.

Çetin, M., & Karsantık, I. (2022). Current trends in school management: School leadership in education 4.0. In Y. Alpaydın & C. Demirli (Eds.), *Educational theory in the 21st century: Science, technology, society and education* (pp. 197–216). Springer.

Chawla, S., & Lenka, U. (2018). Leadership in VUCA environment. In S. Dhir & Sushil (Eds.), *Flexible strategies in VUCA markets* (pp. 213–224). Springer.

Cinnioğlu, H. (2020). A review of modern leadership styles in perspective of industry 4.01. In Akkaya, B. (Eds.), *Agile business leadership methods for industry 4.0* (pp. 1–23). Emerald Publishing Limited.

Coopersmith, K. (2022). Personal development planning and vertical leadership development in a VUCA World. *The Journal of Values-Based Leadership, 15*(1), 1–27. https://doi.org/10.22543/0733.151.1380

Day, D. V., Fleenor, J. W., Atwater, L. E., Sturm, R. E., & McKee, R. A. (2014). Advances in leadership and leader development: A review of 25 years of research and theory. *The Leadership Quarterly, 25*(1), 63–82. https://doi.org/10.1016/j.leaqua.2013.11.004

Delves, R. (2022). Leadership for an unknowable tomorrow. In S. K. Dhiman, & J. F. Marques (Eds.), *Leadership after COVID-19: Working together toward a sustainable future* (pp. 99–114). Springer.

Dhillon, R., & Nguyen, Q. (2020). *Strategies to respond to a VUCA world: A review of existing knowledge.* Department of Business Administration, Lund Universitet.

Dhiman, K. (2022). Holistic leadership for post-COVID-19 organizations:Perspectives and prospects. In S. K. Dhiman, & J. F. Marques (Eds), *Leadership after COVID-19: Working together toward a sustainable future* (pp. 1–16). Springer.

Drucker, P. F. (1980). The deadly sins in public administration. *Public Administration Review, 40*(2), 103–106. https://doi.org/10.2307/975619

Dweck, C. S. (2008). *Mindset: The new psychology of success.* Random House Digital.

Fullan, M. (2021). The right drivers for whole system success. *CSE Leading Education Series, 1.*

Hargreaves, A., & Fink, D. (2006). Redistributed leadership for sustainable professional learning communities. *Journal of School Leadership, 16*(5), 550–565.

Heinonen, S., Karjalainen, J., Ruotsalainen, J., & Steinmüller, K. (2017). Surprise as the new normal: Implications for energy security. *European Journal of Futures Research, 5*(12). https://doi.org/10.1007/s40309-017-0117-5

Horney, N., Pasmore, B., & O'Shea, T. (2010). Leadership agility: A business imperative for a VUCA world. *People & Strategy, 33*(4), 32–38.

Johansen, B. (2012). *Leaders make the future: Ten new leadership skills for an uncertain world.* Berrett-Koehler.

Johansen, B., & Euchner, J. (2013). Navigating the VUCA world. *Research Technology Management, 56*(1), 10–15. https://doi.org/10.5437/08956308x5601003

Kaivo-oja, J. R. L., & Lauraeus, I. T. (2018). The VUCA approach as a solution concept to corporate foresight challenges and global technological disruption. *Foresight, 20*(1), 27–49.

Kok, J., & van den Heuvel, S. C. (2019). *Leading in a VUCA world: Integrating leadership, discernment and spirituality* (p. 209). Springer.

Kukreja, J. (2019). Holacracy: The next generation leadership in a VUCA world. *International Journal of Trend in Scientific Research and Development, 3*(6), 37–47.

Lang S., & Sutton, M. J. D. (2022). Internal–external leadership model in the post-COVID-19, Digital VUCA World. In S. K. Dhiman, & J. F. Marques (Eds), *Leadership after COVID-19: Working together toward a sustainable future* (pp. 471–496). Springer.

Lawrence, K. (2013). *Developing leaders in a VUCA environment* (pp. 1–15). UNC Executive Development.

Li, C., & Lalani, F. (2020). *The COVID-19 pandemic has changed education forever. This is how.* World Economic Forum. https://www.weforum.org/agenda/2020/04/coronavirus-education-global-covid19-online-digital-learning/

Mack, O., & Jungen, M. (2016). Program management in VUCA environments: Theoretical and pragmatical thoughts on a systemic management of projects and programs. In O. M. A. Khare, A. Krämer, & T. Burgartz (Eds.), *Managing in a VUCA World* (pp. 41–57). Springer.

Mahlomotja, R. G. (2018). *Leadership dynamics in a VUCA business world* [Unpublished doctoral dissertation]. University of Pretoria.

Mayer, C. H., & Wilke, K. (2022). Leading through VUCA times for a sustainable future of work: Expert views on the global automotive industry. In S. K. Dhiman, & J. F. Marques (Eds.), *Leadership after COVID-19: Working together toward a sustainable future* (pp. 253–268). Springer.

Millar, C. C. J. M., Groth, O., Mahon, J. F. (2018). Management innovation in the VUCA world: Challenges and recommendations. *California Management Review, 61*(1), 5–14.

Panthalookaran, V. (2022). Education in a VUCA-driven world: Salient features of an entrepreneurial pedagogy. *Higher Education for the Future, 9*(2), 234–249.

Pertiwi, R., Firdaus, M. F., & Sarjito, A. (2022). Leadership challenges in the Vuca (Volatility, Uncertainly, Complexity, Ambiguity) environment. *Jurnal Kewarganegaraan, 6*(2), 2529–2534.

Petrie, N. (2011). Future trends in leadership development. *Center for Creative Leadership White Paper, 5*(5), 36.

Ramakrishnan, D. (2021). Leading in a VUCA world. *Ushus Journal of Business Management, 20*(1), 89–111. https://doi.org/10.12725/ujbm.54.5

Rimita, K. (2019). *Leader readiness in a volatile, uncertain, complex, and ambiguous (VUCA) business environment* [Unpublished doctoral dissertation]. Walden University.

Rimita, K., Hoon, S. N., & Levasseur, R. (2020). Leader *Proceedings of 11th International Conference.* Retrieved from Readiness in a Volatile, Uncertain, Complex, and Ambiguous Business Environment. *Journal of Social Change, 12*(1), 10–18.

Sarkar, A. (2016). We live in a VUCA world: The importance of responsible leadership. *Development and Learning in Organization: An International Journal, 30*(3), 9–12 https://doi.org/10.1108/DLO-07-2015-0062

Sharma, R. (2019). The straits of success in a VUCA world. *ABS International Journal of Management, 7*(1), 1–6.

Sinha, D., & Sinha, S. (2020). Managing in a VUCA world: Possibilities and pitfalls. *Journal of Technology Management for Growing Economies, 11*(1), 17–21. https://doi.org/10.15415/jtmge.2020.111003

Smith, L., & Riley, D. (2012). School leadership in times of crisis. *School Leadership & Management, 32*(1), 57–71. https://doi.org/10.1080/13632434.2011.614941

Sullivan, J. (2012, January 16). *VUCA: The new normal for talent management and workforce planning.* Ere.net. Retrieved from http://www.ere.net/2012/01/16/vuca-the-new-normal-for-talent-management-and-workforce-planning/

Suzana, A. L., & Mohd Amir Shaukhi, A. (2020). Learning agility among educational leaders: A Vuca-ready leadership competency? *Jurnal Pengurusan Dan Kepimpinan Pendidikan, 33*(1).

Suzana, A. L., & Sazali, Y. (2013). *Leadership development for educational leaders; An integrated model.* Institut Aminuddin Baki.

Tadesse, S., & Muluye, W. (2020). The impact of COVID-19 pandemic on education system in developing countries: A review. *Open Journal of Social Sciences, 8*(10), 159–170. https://doi.org/10.4236/jss.2020.810011

Thebe, S., & Taran, Y. (2017). *Managing change in a VUCA world a retrospective case studies of Ford and GM* [Unpublished doctoral dissertation]. Aalborg University.

White, J. M. (2022). *Effective educational leadership traits when transitioning between K-12 environments in a VUCA world* [Unpublished doctoral dissertation]. University of Charleston].

Yurdasever, E., & Fidan, Y. (2020). KOMB (VUCA) world and new leadership skills. *Journal of the Human and Social Sciences Research, 9*(2), 1638–1664.

Part 2: Leading Change: Skills to Help Educational Leaders Thrive in a VUCA World

Dr. Özge Öztekin Bayır

Unveiling the Role of Authentic Leadership for Educational Leaders in a VUCA World

Abstract
The world of today is a fully VUCA environment, where it is impossible to predict the speed and extent of change and where COVID-19 makes planning and forecasting challenging. In order to successfully handle the challenges of this VUCA environment, a different kind of leadership is needed in every field, including education. Educational leaders must build the necessary set of skills to function thoughtfully and resolutely in the face of this constantly increasing volatility, uncertainty, complexity, and ambiguity (New VUCA), as basic abilities are insufficient to handle these challenges. Therefore, in today's world, authenticity holds greater significance compared to charisma and power. Authentic leadership is founded on moral principles that are value-driven. In these shifting times, authentic leaders with their facilitating role have potential to inspire their people to adopt new viewpoints, question established ideas, formulate more empowering concepts, and cultivate flexibility, inventiveness, and self-assurance. Also, educational leaders should think and act differently in order to reduce the direct and indirect damages related the VUCA environment. If modern education leaders possess the authentic leadership competencies that connect the body, mind and spirit, they can ensure a shared and sustainable future, and demonstrate leadership in the emerging VUCA environment at schools. In light of all these relevant information, this chapter aims to examine the role of authentic leadership which serves as the foundation for various other positive leadership styles in the current VUCA environment at schools. This chapter also helps educational leaders to equip them to function effectively and efficiently in the modern educational systems that have faced many challenges.

Keywords: Authentic leadership, authenticity, VUCA, educational leadership in changing times

> *"Some magic takes places in the crucible of leadership."*
>
> *—Warren Bennis, On Becoming a Leader*

Educational organizations strive to adapt to the rapidly changing world conditions in recent years. One of these changing conditions is the VUCA world. The word "VUCA" stands for "volatility, uncertainty, complexity, and ambiguity" and emphasizes the things which are uncertain, complex, and difficult to understand. The rapid change in the structure of business environments, technological developments and global events have led to the widespread use of this term in recent years. In this changing world, leadership has gained a different dimension. Now, leaders should not only manage a business, but also have leadership skills that can quickly adapt to changing conditions and lead their organizations to success. In this context, authentic leadership has become an important leadership approach for educational organizations.

Authentic leadership focuses on the leader knowing himself, sticking to his values, and setting an example for his employees. By establishing open and honest communication, the leader gives confidence to his employees and helps to increase their performance (Chan, 2005). In addition, authentic leaders create a common vision by acting together with their employees while managing the change processes in their organizations (Avolio & Gardner, 2005; Walumbwa et al., 2008). In addition, authentic leaders give confidence to their employees (Feng, 2016; Luthans & Avolio, 2003), create a common vision and lead their organizations to success by acting together with their employees in changing processes. In this context, educational organizations can adopt an authentic leadership approach to ensure the development of students, prepare them for the future and adapt quickly to changing conditions in the VUCA world.

Leadership

Since the development of the social sciences, the study of leadership has evolved into a scholarly field of interest for a variety of academic disciplines, including sociology, psychology, political science, and others. Due to the complexity and multifaceted nature of the concept of leadership, scholars in the social sciences have put forth various ideas to try and make sense of leadership and how it manifests at organizational level (Arar & Oplatka, 2022). There are different perspectives and approaches to defining leadership, depending on the context and the discipline. The concept of leadership, which began to be studied by scientists in the early twentieth century, was analyzed with innate qualities and characteristics (Stogdill, 1948), actions and behaviors (Blake & Mouton, 1964), situational factors (Hersey et al., 2008), inspiration, motivation and vision (Bass, 1990), transparency, ethical behavior and serving the needs of the followers (Avolio & Gardner, 2005; Greenleaf, 1977). Thus, concept of leadership can be defined in many ways.

Leadership is generally defined as the process of guiding, organizing and controlling the behavior of group members (Kotter, 1990; Yukl, 2013). Northouse (2021), who states that some scholars conceptualize leadership as a trait or a set of behaviors, whereas others adopt a perspective that focuses on information processing or the formation of relationships when examining the phenomenon, defines leadership as "a process whereby influences a group of individuals to achieve a common goal." In his definition, there are four dimensions as critical to the phenomenon of leadership. These are the (i) process dimension of leadership that emphasizes the communication between the leader and his followers rather than just a characteristic feature, (ii) the influence dimension that states that the leader influences his followers, (iii) the group dimension that draws attention to the fact that leadership takes place in groups with common goals, and (iv) the goal attainment dimension, which includes the leaders' efforts for their followers to reach certain goals (Northouse, 2021). In general, these leadership theories and definitions offer many viewpoints on leadership, emphasizing various facets of effective leadership and proposing various methods for leaders to inspire and encourage followers.

Leading in a VUCA World: The Need for Different Leadership

The rapid advancement of technology, coupled with a significant rise in population, influx of immigrants, catastrophic ecological events, and shifting geopolitical landscapes, continue to bring about profound changes and disruptions to our daily lives. These developments are further amplified by the global interconnectedness brought about by globalization. The overall effect of these complex and dynamic factors is often referred to as VUCA (Hadar et al., 2020). A world that is volatile, unpredictable, complex, ambiguous (VUCA) has the following characteristics (Horney et al., 2010):

(i) Volatility: The nature, speed, volume, magnitude and dynamics of change,
(ii) Uncertainty: The lack of predictability of issues and events,
(iii) Complexity: The confounding of issues and the chaos that surround any organization,
(iv) Ambiguity: The haziness of reality and the mixed meanings of conditions.

VUCA is an abbreviation that describes the conditions and factors that are encountered in almost every business today. It also refers to an environment that demands you to adapt immediately, take appropriate action in unpredictable circumstances (Yurdaseven & Fidan, 2020). Watt et al. (2017) indicate that it is imperative that VUCA and its impacts on organizational procedures be considered during the decision-making process as well as the training of leaders for the twenty-first century. So, educational leaders as well as other leaders at the executive level need to comprehend VUCA situations, anticipate their consequences, and be prepared for many crises characterized by volatility, uncertainty, complexity, and ambiguity. As pointed out by Chawla and Lenka (2018), in constantly shifting environments, leaders are required to generate novel concepts and actively involve their people in fresh starts. The term VUCA world including volatile, uncertain, complex and ambiguous situations also defines some specific difficulties that organizations must overcome in order to assist leaders in making progress with the newly accessible options and opportunities.

Based on the view that in times of change, leaders need to encourage the people they lead to introduce new points of view, question long-held beliefs, think of more powerful ideas, build tolerance, inventiveness, and self-assurance, and produce more empowered attitudes (Chawla & Lenka, 2018), using a variety of approaches to leadership may bring a wide range of helpful viewpoints to be able to make it through this VUCA environment. The COVID-19 pandemic is just one of the situations in the current VUCA setting and it brings new difficulties include a loss of familiarity, chaos, unpredictability in the pace and scope of change, abrupt and serious fluctuations brought on by lockdowns (Kumar & Modi, 2022). The COVID-19 pandemic has highlighted the VUCA characteristics of the world (Mielkov et al., 2021) and has affected various aspects of the educational, organizational and communication systems. The pandemic has forced these systems to adapt and transform to thrive in the VUCA era.

The pandemic has brought to light the concept of a VUCA world—the term used to describe an environment characterized by volatility, uncertainty, complexity, and ambiguity. In the context of education, the pandemic has introduced significant challenges for educators and students, requiring them to be agile and adaptable in the face of unexpected and rapidly changing circumstances (Aristovnik et al., 2020; Pokhrel & Chhetri, 2021; Tadesse & Muluye, 2020; Tarkar, 2020). It is possible to consider the effects of the four distinct qualities (volatility, uncertainty, complexity, and ambiguity) that define the world of VUCA in educational organizations during the COVID-19 pandemic as follows:

Volatility in the context of education refers to the unexpected and quick changes brought on by the pandemic. For schools and universities, it has been an unpredictable experience as they have been forced to either close or switch to remote learning (Stanistreet et al., 2020). Both teachers and pupils experienced a wave of confusion and instability as a result of this sudden shift (Tarkar, 2020). Teachers had to quickly adopt new teaching strategies, accept novel technologies, and look at different approaches to determining student success. Meanwhile, students faced the challenges of adapting to different learning environments and developing new routines (Camacho-Zuñiga et al., 2021; Jena, 2020). The dynamic educational environment required adaptability and flexibility which served as a reminder of the human resilience and determination that exist at the core of education.

Uncertainty within the educational environment denotes the absence of predictability and the dearth of information regarding the forthcoming developments in the realm of education. The concept of uncertainty is also intrinsically linked to individuals' lack of comprehension regarding the current state of affairs in actuality (Bennett & Lemoine, 2014). According to Jung et al. (2021), uncertainty is the pandemic's most significant residual effect, affecting society as a whole and the pandemic has introduced significant uncertainty into the education sector. This situation has made it challenging for teachers, academics and students to plan for the future and has created anxiety and stress among them (Ardan et al., 2020; Burak, 2023; Pressley, 2021; Pressley et al., 2021).

Complexity describes a scenario with numerous interconnected factors that is typically challenging to manage (Bennett & Lemoine, 2014). When we associate the concept of complexity with education, it refers to the intricate and interrelated systems that make it challenging to understand and manage. The pandemic has made it clear that education is a complex system. This intricate process that entails swift alterations in education and training procedures, thereby leaving the stakeholders unprepared for such unexpected changes, necessitated the principals to lead the disruption in education by promptly modifying their strategies and approaches (Dare & Saleem, 2022).

Ambiguity has been characterized as a lack of clarity surrounding an event or the ambiguous and challenging causes behind the things occurring (Waller et al., 2019). With contradicting facts and perceptions of the virus and its effects, the pandemic has significantly increased ambiguity in the educational field. As a result, teachers and students have found it challenging to navigate the situation and react appropriately (Pokhrel & Chhetri, 2021).

The COVID-19 pandemic has been classified as a black swan occurrence by numerous scholars owing to its unforeseen and severe repercussions on societies and economies around the world (Devarajan et al., 2021; Dogra & Parrey, 2022; Hadar et al., 2020). According to Taleb (2007), a black swan event is characterized as "an event that is highly unpredictable, has a massive impact, and is only rationalized in hindsight." While the pandemic has caused widespread disruption, it has also presented opportunities for innovation and change in various areas, such as education (Stanistreet et al., 2020). It is important to identify and capitalize on these opportunities to create a better, more resilient education systems. The education sector has been profoundly affected by the COVID-19 pandemic, leading to the closure or transition to remote learning of schools and universities worldwide.

The pandemic process has caused a paradigm shift in every field (Howe et al., 2021). To reduce the damages that COVID-19 cause, leaders in all fields like political, social, educational and business have to adopt new ways of thinking and leading. Especially with the pandemic period, managing remote teams, leading change, and coping with uncertainty have become more important for educational leaders.

Authentic Leadership

A VUCA environment is one that is continually changing and enduring situations that regularly test a leader's capacity to make decisions, create strategies, and carry them out perfectly (Bernstein, 2014). Therefore, there has been a boom in research on whether or not today's education leaders have the necessary leadership abilities to enable them to successfully and efficiently navigate the roller-coaster scenarios that threaten today's educational systems (Mahel, 2021). Leaders need to continually develop and adjust their abilities to align with the social and situational conditions of their diverse roles, assignments, and organizations in the face of complex, global, competitive, unpredictable, volatile, and ambiguous environments and circumstances (Littlejohn, 2019). Leaders will need to grasp complicated dynamics in order to identify that a collaborative process must be distributed throughout people networks (Rodriguez & Rodriguez, 2013). Authentic leadership, on the other hand, can assist leaders in coping with the challenges they encounter in a VUCA world.

Authentic leadership focuses on leaders' self-awareness, staying true to their values, and acting with honesty. Avolio et al. (2004) define authentic leadership as a leadership style exhibited by individuals who are aware of their own values and thoughts, self-confident, optimistic, and value the emotions and thoughts of their followers, and possess high ethical standards. Authentic leadership is a leadership approach that focuses on the leader's self-awareness, transparency, ethical behavior, and values-based decision-making (Walumbwa et al., 2008). It is characterized by the leader's ability to express genuine and authentic emotions, create a positive organizational climate, and build trusting relationships with followers (Avolio & Gardner, 2005).

According to Avolio and Gardner (2005), authentic leadership is a fundamental concept and due to this feature, authentic leadership forms the basis of other positive leadership approaches. Authentic leadership is often associated with concepts from positive psychology, such as hope, trust and flexibility. Authentic leaders, by virtue of possessing these positive competencies, have a multitude of positive impacts on organizational processes (Hassan & Ahmed, 2011).

There is empirical support indicating that adopting an authentic leadership approach is effective in generating successful and sustainable outcomes within organizations (George, 2003). For instance, the authentic leadership approach has been discovered to hold a pivotal function in reducing employee burnout within organizational settings (Adil & Kamal, 2018). In challenging circumstances, it is extremely crucial for any organization to have leaders who are optimistic and hopeful. At this point, the authentic leadership approach provides a new perspective. This perspective stresses the significance of leaders who are not just charismatic or transformative (inspiring their followers with new ideas and personal care and encouraging their own leadership growth), but also self-assured, hopeful, optimistic, and resilient. Such leaders are capable of influencing their followers to adopt the same optimistic outlook as they do because they exhibit behaviors that show confidence, hope, optimism, resilience, transparency, and integrity. Through this mechanism, leaders' enthusiasm and honesty spread to their followers. It's very clear that optimistic organizational leaders are required during these challenging times so that people might restore a sense of control over their own lives and be encouraged to seek out new routes (Youssef & Luthans, 2006). The concept of authentic leadership primarily concerns the authenticity and identity of the leader (Chan, 2005; Northouse, 2021). Therefore, in order to fully comprehend the concept of authentic leadership, a more detailed examination of what authenticity entails is required.

Authenticity

The amount of research on authentic leadership has significantly risen during the past 10 years (Alilyyani et al., 2018). Since it is a novel idea in the leadership literature, authentic leadership has recently been discussed in modern leadership theories. Öztekin-Bayır (2018) indicated that it is important to concentrate on what the word "authenticity" implies in various settings in order to make the concept of authentic leadership more clear.

The origins of the idea of authenticity can be found in Greek philosophy, which emphasized discovering one's actual self (Karadağ & Öztekin-Bayır, 2018). Authenticity entails acting in accordance with one's genuine self, expressing oneself in ways that are congruent with inner ideas and feelings (Harter, 2002). Authenticity, according to this viewpoint, is plainly a property of whatever which claims to be authentic and arises as a natural or spontaneous phenomenon (Iszatt-White & Kempster, 2018). Therefore, being authentic as a person involves being truthful to oneself and others in a way that a person's words and actions are consistent. A true leader, then, exhibits consistency in both his actions and his feelings and thoughts (Novicevic et al., 2006).

According to Ilies et al. (2005), authenticity refers to a wide psychological concept that encompasses a person's propensity to see themselves within their social context and to lead their lives in accordance with their core beliefs. In addition to this view, it is stated that authenticity promotes consistency in the trinity of emotion, thought, and behavior (Gardner et al., 2005). With this consistency, authenticity enables a person to openly communicate their thoughts, feelings, and wishes; it has an impact on both the individual and others around him (Klenke, 2007).

The Components of Authentic Leadership

According to Onyalla (2018), components of authentic leadership are important to other leadership ideas, and instead of analyzing authentic leadership as a separate construct, understanding this style in cooperation with other leadership styles would be beneficial. Being fairly broad, the word "authentic leadership" "may contain transformational, charismatic, servant, spiritual, or other styles of positive leadership" (Avolio & Gardner, 2005). Authentic leaders are conscious of the environment in which they work, and they also possess strong moral principles and are self-assured, positive, and optimistic. According to Walumbwa et al. (2008) there are four crucial characteristics of authentic leaders: (*i*) *self-awareness*, (*ii*) *balanced processing*, (*iii*) *internalized moral perspective* and (*iv*) *relational transparency*.

Self-awareness

According to Gardner et al. (2005), authentic leadership development relies heavily on the cultivation of self-awareness. This implies that individuals aspiring to be authentic leaders must engage in introspection and self-reflection to gain a profound comprehension of their values, beliefs, strengths, weaknesses, and biases. By attaining such a level of self-awareness, individuals can foster the authenticity necessary to effectively lead others. In the words of Walumbwa et al. (2008), authentic leaders with heightened self-awareness demonstrate sensitivity toward how their followers are influenced by their own strengths and weaknesses. This sensitivity arises from their self-awareness, as they possess a deep understanding of their abilities, tendencies, and behaviors, which they leverage to lead and influence their followers effectively (Ilies et al., 2005).

The notion of self-awareness revolves around developing a comprehensive understanding of oneself—encompassing one's personal strengths and the areas for development along with perceiving the diverse facets of emotions associated with daily experiences. This process plays a pivotal role in authentic leadership where leaders need to recognize the impact that their distinct personalities have on those they lead (Kernis, 2003).

Balanced Processing

The process entails embracing an objective and accepting perspective concerning one's positive and negative attributes, qualities, and characteristics, while avoiding negation, misrepresentation, amplification, or disregard of personal cognizance, inner perceptions, and externally sourced evaluative data (Kernis, 2003). Avolio (2007) employs the term "balanced processing" to refer to leaders' capacity to objectively analyze and evaluate all relevant information pertinent to a given situation prior to decision-making. This necessitates a fair and impartial appraisal of all available data, without being influenced by personal biases, emotions, or preconceived notions (Walumbwa et al., 2008). It alludes to leaders' inclination and readiness to explore the concepts of others before reaching a decision. Authentic leaders are receptive to the perspectives of their subordinates and are open to various viewpoints. Consequently, they adopt an objective evaluation process (Gardner et al., 2005).

Internalized Moral Perspective

The concept of internalized moral perspective revolves around how an individual's own moral values shape their behavior. Within the realm of authentic leadership, leaders are revered for their authenticity, as they exhibit a remarkable alignment between their actions and their deeply ingrained ethical values and principles. Authentic leaders place great importance on their ethical compass, utilizing it as a guiding force to shape their behaviors in accordance with their fundamental convictions (Avolio & Gardner, 2005). This authentic approach not only cultivates trust and loyalty among their followers but also enhances their credibility as leaders who genuinely embody their professed values. By staying true to themselves and their principles, authentic leaders forge meaningful connections with those they lead, fostering an environment of trust, respect, and integrity. Moreover, adopting this perspective encourages leaders to take ownership and be accountable for their actions, recognizing the impact they have on others. In simpler terms, authentic leaders lead with integrity, staying true to their values, and inspiring others to do the same (Kernis, 2003; Luthans & Avolio, 2003).

Relational Transparency

The idea of relational transparency, as described by Gardner et al. (2005), is about authentic leaders effectively expressing their personal beliefs and values to their team members, demonstrating an authentic sense of awareness. It highlights the importance of clear and direct communication in the relationships leaders build. According to Luthans and Avolio (2003), an essential element of authentic leadership is being relationally transparent, which means fostering an environment where information is openly and honestly shared. In simpler terms, authentic leaders believe in being transparent in their relationships, which helps create trust and openness within the organization. They not only encourage honesty and openness themselves but also inspire their followers to do the same, establishing a culture of trust and transparency (Avolio et al., 2009; Ilies et al., 2005; Walumbwa et al., 2008).

What Authentic Leadership Tells Educational Leaders in a VUCA World?

Authentic educational leadership has become a prominent focus in educational research and practice, highlighting the possibility of leaders embodying their true selves, displaying authenticity, and fostering a culture of positivity and ethics within organizations. Educational leaders operate in a constantly changing and challenging world filled with unpredictability, complexity, and ambiguity. In such a VUCA (Volatility, Uncertainty, Complexity, and Ambiguity) environment, authentic leadership becomes an invaluable compass that guides the stakeholders. Authentic leadership offers them valuable insights and guidance to tackle the unique challenges they face.

It is possible to examine the different perspectives that the authentic leadership approach can offer to educational leaders within the following dimensions:

(i) *Perspective centered on morality and values:* The prominence of authentic leadership stems from its emphasis on leaders who prioritize transparency and possess a robust ethical dimension (Dhiman, 2017). People who follow authentic leaders see them as positive role models because they consistently act with integrity, strongly uphold ethical values, and foster a supportive organizational environment (Klenke, 2007). As described by Luthans and Avolio (2003), authentic leaders are motivated by a core set of values that prioritize doing what is morally right and just for the individuals and communities they lead. For instance, it is crucial for educational leaders to make impartial and values-driven decisions that consider the well-being of both students and teachers. By doing so, they avoid decisions that solely serve their own interests and instead foster a positive and constructive culture within the school (Karadağ & Öztekin-Bayır, 2018). This can be achieved by establishing a decision-making framework rooted in principles of fairness, ethics, and justice, ultimately contributing to the development of a positive and inclusive educational environment.

(ii) *Building trust:* There also exists a positive association between authentic leadership and employee trust, specifically through the leader's consistency in aligning their words and actions (Alilyyani et al., 2018). In other words, when leaders consistently demonstrate congruence between what they say and what they do, it fosters trust among their employees. Avolio et al. (2004) indicate that authentic leaders foster a sense of hope in their followers and establish trust in leadership within their follower base. As stated by Tschannen-Moran (2004), a sense of trust among organizational members can contribute to the formation of a positive organizational culture. Additionally, it has been noted that schools with high levels of trust exhibit improvements in student learning outcomes (Seashore-Louis et al., 2010). Authentic leaders are also real individuals who stay true to themselves and their convictions. They form authentic relationships with their staff, which builds trust. This trust allows educational leaders to inspire and motivate teachers, leading to higher levels of performance (Arar & Oplatka, 2022).

(iii) *Transparency:* School leaders aspire to demonstrate honesty and integrity, openly share their perspectives, operate with trustworthiness and transparency, and exhibit both intellectual acuity and discernment in their leadership of teachers, students, and the wider school community (Alazmi & Al-Mahdy, 2022). Integrity and transparency ensure that the school maintains effective leadership even in crisis situations or uncertainties. Therefore, it is important for school leaders to regularly communicate with stakeholders, providing clear information about the school's goals, policies to be implemented, and decisions to be made in order to ensure transparency. Furthermore, considering the contribution of transparency in creating a healthier and more optimistic work environment within the organization (Alilyyani et al., 2018), it becomes evident why school leaders should adopt an open and honest communication approach in the school setting.

In a VUCA world, authentic leadership emphasizes the importance of transparent communication. Educational leaders need to openly share information, clarify expectations, and provide a clear sense of direction to mitigate uncertainties. Transparent communication fosters trust, reduces ambiguity, and allows stakeholders to make informed decisions (Avolio et al., 2004).

(iv) *Strong communication skills:* Authentic leadership encourages educational leaders to nurture positive relationships and create a supportive organizational climate (Walumbwa et al., 2008). It is important for school leaders to have effective communication skills as it enables information flow and collaboration within the school. Therefore, school leaders should engage in open and effective communication with students and teachers, actively listening to them and providing regular feedback. Building strong connections, fostering collaboration, and promoting a sense of belonging among team members contribute to resilience and adaptability in a VUCA environment.

(v) *Value-based and ethically grounded decisions:* Authentic leadership places a strong emphasis on the ethical dimension of decision-making (Cianci et al., 2014). In the ever-evolving and unpredictable educational landscape we find ourselves in today, leaders are confronted with intricate challenges and hazy circumstances that demand not only astute discernment but also unwavering integrity. These educational leaders navigate through a maze of complexities, weighing the moral implications of their choices and striving to uphold ethical standards. They recognize the significance of balancing short-term obstacles with long-term educational goals, always guided by a steadfast moral compass (Gardner et al., 2005). Educational leaders may build strong relationships with their staff members and stakeholders by adopting this ethical framework, which also fosters effective communication. Additionally, they help create a collaborative, motivated, and learning-focused culture inside the educational setting.

Conclusion

Authentic leadership holds great significance for educational leaders in the VUCA world. This leadership approach represents leaders who are true to themselves and their values, ensuring alignment between their personalities and beliefs. They prioritize staying authentic and genuine in their actions and decisions. By embracing the authentic leadership approach, school leaders foster effective communication with teachers and students, which leads to the development of a trusting environment in the school. They actively strive to understand the needs of both teachers and students, offering support and making transparent decisions. This commitment to transparency and support contributes to the formation of a positive school culture. As a result, the increased motivation levels within the school positively impact learning outcomes, creating a conducive environment for student success. In conclusion, by prioritizing the growth and development of their team members and creating and developing a motivated and learning-oriented culture within the school atmosphere where everyone feels valued and supported, authentic leadership plays a critical role in bringing about positive change and transformation in the field of education within the VUCA world.

References

Adil, A., & Kamal, A. (2018). Impact of perceived authentic leadership and psychological capital on burnout: Mediating role of psychological ownership. *Psychological Studies, 63*(3), 243–252.

Alazmi, A. A., & Al-Mahdy, Y. F. H. (2022). Principal authentic leadership and teacher engagement in Kuwait's educational reform context. *Educational Management Administration & Leadership, 50*(3), 392–412.

Alilyyani, B., Wong, C. A., & Cummings, G. (2018). Antecedents, mediators, and outcomes of authentic leadership in health care: A systematic review. *International Journal of Nursing Studies, 83*, 34–64.

Arar, K., & Oplatka, I. (2022). *Advanced theories of educational leadership*. Springer.

Ardan, M., Rahman, F. F., & Geroda, G. B. (2020). The influence of physical distance to student anxiety on COVID-19, Indonesia. *Journal of Critical Reviews, 7*(17), 1126–1132.

Aristovnik, A., Keržič, D., Ravšelj, D., Tomaževič, N., & Umek, L. (2020). Impacts of the COVID-19 pandemic on life of higher education students: A global perspective. *Sustainability, 12*(20), 8438. https://doi.org/10.3390/su12208438

Avolio, B. J. (2007). Promoting more integrative strategies for leadership theory-building. *American Psychologist, 62*(1), 25–33.

Avolio, B. J., & Gardner, W. L. (2005). Authentic leadership development: Getting to the root of positive forms of leadership. *The Leadership Quarterly, 16*(3), 315–338.

Avolio, B. J., Gardner, W. L., Walumbwa, F. O., Luthans, F., & May, D. R. (2004). Unlocking the mask: A look at the process by which authentic leaders impact follower attitudes and behaviors. *The Leadership Quarterly, 15*(6), 801–823.

Avolio, B. J., Walumbwa, F. O., & Weber, T. J. (2009). Leadership: Current theories, research, and future directions. *Annual Review of Psychology, 60*, 421–449.

Bass, B. M. (1990). *Bass and Stogdill's handbook of leadership: Theory, research & managerial applications*. Free Pres.

Bennett, N., & Lemoine, G. J. (2014). What a difference a word makes: Understanding threats to performance in a VUCA world. *Business Horizons*, *57*(3), 311–317.

Bennis, W. (2009). *On becoming a leader*. Basic Books.

Bernstein, L. E. (2014). *The perceived importance of Vuca-driven skills for 21st century leader success and the extent of integration of those skills into leadership development programs* [Doctoral dissertation, Drake University].

Blake, R. R., & Mouton, J. S. (1964). *The managerial grid*. Gulf.

Burak, D. (2023). The effect of risk and protective factors on primary school students' COVID-19 anxiety: Back to school after the pandemic. *Child Indicators Research*, *16*(1), 29–51.

Camacho-Zuñiga, C., Pego, L., Escamilla, J., & Hosseini, S. (2021). The impact of the COVID-19 pandemic on students' feelings at high school, undergraduate, and postgraduate levels. *Heliyon*, *7*(3). https://doi.org/10.1016/j.heliyon.2021.e06465

Chan, A. (2005). Authentic leadership measurement and development: Challenges and suggestions. In W. L. Gardner, B. J. Avolio, & F. O. Walumbwa (Eds.), *Authentic leadership theory and practice: Origins, effects, and development* (pp. 227–251). Elsevier Science.

Chawla, S., & Lenka, U. (2018). Leadership in VUCA environment. In S. Dhir & Sushil (Eds.), *Flexible strategies in VUCA markets* (pp. 213–224). Springer.

Cianci, A. M., Hannah, S. T., Roberts, R. P., & Tsakumis, G. T. (2014). The effects of authentic leadership on followers' ethical decision-making in the face of temptation: An experimental study. *The Leadership Quarterly*, *25*(3), 581–594. https://doi.org/10.1016/J.LEAQUA.2013.12.001

Dare, P. S., & Saleem, A. (2022). Principal leadership role in response to the pandemic impact on school process. *Frontiers in Psychology*, *13*, 943442. https://doi.org/10.3389/fpsyg.2022.943442

Devarajan, J. P., Manimuthu, A., & Sreedharan, V. R. (2021). Healthcare operations and black swan event for COVID-19 pandemic: A predictive analytics. *IEEE Transactions on Engineering Management*. https://doi.org/10.1109/TEM. 2021.3076603

Dhiman, S. (2017). Leadership and spirituality In J. Marques & S. Dhiman (Eds.), *Leadership today: Practices for personal and professional performance* (pp. 139–160). Springer.

Dogra, P., & Parrey, A. H. (2022). Work from home amid black swan event (Covid-19): A bibliometric analysis from a social science perspective. *Kybernetes*. https://www.emerald.com/insight/0368-492X.htm

Feng, F. I. (2016). School principals' authentic leadership and teachers' psychological capital: Teachers' perspectives. *International Education Studies*, *9*(10), 245–255.

Gardner, W. L., Avolio, B. J., Luthans, F., May, D. R., & Walumbwa, F. O. (2005). "Can you see the real me?" A self-based model of authentic leader and follower development. *The Leadership Quarterly*, *16*, 343–372.

George, W. (2003). *Authentic leadership: Rediscovering the secrets to creating lasting value*. Jossey-Bass.

Greenleaf, R. K. (1977). *Servant leadership: A journey into the nature of legitimate power and greatness*. Paulist Press.

Hadar, L. L., Ergas, O., Alpert, B., & Ariav, T. (2020). Rethinking teacher education in a VUCA world: Student teachers' social-emotional competencies during the Covid-19 crisis. *European Journal of Teacher Education*, *43*(4), 573–586.

Harter, S. (2002). Authenticity. In C. R. Snyder & S. Lopez (Eds.), *Handbook of positive psychology* (pp. 382–394). Oxford University Press.

Hassan, A., & Ahmed, F. (2011). Authentic leadership, trust and work engagement. *International Journal of Human and Social Sciences*, *6*(3), 164–170.

Hersey, P., Blanchard, K. H., & Johnson, D. E. (2015). *Management of organizational behavior: Leading human resources*. Pearson.

Horney, N., Pasmore, B. & O'Shea, T. (2010). Leadership agility: A business imperative for a VUCA world. *People & Strategy*, *33*(4), 32–38.

Howe, D. C., Chauhan, R. S., Soderberg, A. T., & Buckley, M. R. (2021). Paradigm shifts caused by the COVID-19 pandemic. *Organizational Dynamics*, *50*(4), 1–9.

Ilies, R., Morgeson, F. P., & Nahrgang, J. D. (2005). Authentic leadership and eudaemonic well-being: Understanding leader–follower outcomes. *The Leadership Quarterly*, *16*(3), 373–394.

Iszatt-White, M., & Kempster, S. (2019). Authentic leadership: Getting back to the roots of the "root construct"? *International Journal of Management Reviews*, *21*, 356–369.

Jena, P. K. (2020). Impact of pandemic COVID-19 on education in India. *International Journal of Current Research*, *12*(7), 12582–12586. https://doi.org/10.24941/ijcr.39209.07.2020

Jung, J., Horta, H., & Postiglione, G. A. (2021). Living in uncertainty: The COVID-19 pandemic and higher education in Hong Kong. *Studies in Higher Education*, *46*(1), 107–120.

Karadağ, E. & Öztekin-Bayır, Ö. (2018). The effect of authentic leadership on school culture: A structural equation model. *International Journal of Educational Leadership and Management*, *6*(1), 40–75. https://doi.org/10.17853/ijelm.2018.2858

Kernis, M. H. (2003). Toward a conceptualization of optimal self-esteem. *Psychological Inquiry*, *14*(1), 1–26.

Klenke, K. (2007) Authentic leadership: A self, leader and spiritual identity perspective. *International Journal of Leadership Studies*, *3*(1), 68–97.

Kotter, J. P. (1990). *General managers*. Simon and Schuster.

Kumar, V., & Modi, S. (2022). Leading in new VUCA environment: Role of positive leadership through spiritual and ethical values. In S. K. Dhiman & J. F. Marques (Eds.), *Leadership after COVID-19: Working together toward a sustainable future* (pp. 561–573). Springer.

Littlejohn, R. (2019). Leadership and diversity implementation. In B. K. Mickahail & C. T. E. Aquino (Eds.), *Effective and creative leadership in diverse workforces: Improving organizational performance and culture in the workplace* (pp. 43–64). Springer.

Luthans, F., & Avolio, B. J. (2003). Authentic leadership development. *Positive Organizational Scholarship*, *241*(258), 1–26.

Mahel, A. (2021). Leadership competencies: Tech-saviness, risk and conflict management challenges to higher education in a volatile, uncertain, complex, and ambiguous (VUCA) environment. *American Journal of Humanities and Social Science (AJHSS)*, *27*, 40–41.

Mielkov, Y., Bakhov, I., Bilyakovska, O., Kostenko, L., & Nych, T. (2021). Higher education strategies for the 21st century: Philosophical foundations and the humanist approach. *Revista Tempos e Espaços em Educação*, *14*(33), 15524. http://dx.doi.org/10.20952/revtee.v14i33.15524

Northouse, P. G. (2021). *Leadership: Theory and practice*. Sage Publications.

Novicevic, M. M., Harvey, M. G., Buckley, M. R., Brown, J. A., & Evans, R. (2006). Authentic leadership: A historical perspective. *Journal of Leadership and Organizational Studies*, *13*(1), 64–76.

Onyalla, D. B. (2018). Authentic leadership and leadership ethics: Proposing a new perspective. *The Journal of Values-Based Leadership*, *11*(2). https://doi.org/10.22543/0733.62.1226

Öztekin-Bayır, Ö. (2018). *The relationship between authentic leadership, psychological capital, work related flow and school achievement* [Doctoral dissertation]. Eskişehir Osmangazi University.

Pokhrel, S., & Chhetri, R. (2021). A literature review on impact of COVID-19 pandemic on teaching and learning. *Higher Education for the Future, 8*(1), 133–141.

Pressley, T. (2021). Factors contributing to teacher burnout during COVID-19. *Educational Researcher, 50*(5), 325–327.

Pressley, T., Ha, C., & Learn, E. (2021). Teacher stress and anxiety during COVID-19: An empirical study. *School Psychology, 36*(5), 367–376.

Rodriguez, A., & Rodriguez, Y. (2015). Metaphors for today's leadership: VUCA world, millennial and "Cloud Leaders." *Journal of Management Development, 34*(7), 854–866.

Seashore Louis, K., Dretzke, B., & Wahlstrom, K. (2010). How does leadership affect student achievement? Results from a national US survey. *School effectiveness and School Improvement, 21*(3), 315–336.

Stanistreet, P., Elfert, M., & Atchoarena, D. (2020). Education in the age of COVID-19: Understanding the consequences. *International Review of Education, 66*, 627–633.

Stogdill, R. (1948). Personal factors associated with leadership: A survey of the literature. *Journal of Psychology, 25*, 35–71.

Tadesse, S., & Muluye, W. (2020). The impact of COVID-19 pandemic on education system in developing countries: A review. *Open Journal of Social Sciences, 8*(10), 159–170.

Taleb, N. N. (2007). Black swans and the domains of statistics. *The American Statistician, 61*(3), 198–200.

Tarkar, P. (2020). Impact of COVID-19 pandemic on education system. *International Journal of Advanced Science and Technology, 29*(9), 3812–3814.

Tschannen-Moran, M. (2004). *Trust matters: Leadership for successful schools.* Jossey Bass.

Waller, R. E., Lemoine, P. A., Mense, E. G., Garretson, C. J., & Richardson, M. D. (2019). Global higher education in a VUCA world: Concerns and projections. *Journal of Education and Development, 3*(2), 73–83.

Walumbwa, F. O., Avolio, B. J., Gardner, W. L., Wernsing, T. S., & Peterson, S. J. (2008). Authentic leadership: Development and validation of a theory-based measure. *Journal of Management, 34*(1), 89–126.

Watt, S., Javidi, M., & Normore, A. H. (2017). Increasing darkness: Combining toxic leadership and volatility, uncertainty, complexity and ambiguity (VUCA). In A. H. Normore & J. S. Brooks (Eds.), *The dark side of leadership: Identifying and overcoming unethical practice in organizations* (pp. 195–206). Emerald Group Publishing.

Youssef, C. M., & Luthans, F. (2005). Positivity in the Middle East: Developing hopeful Egyptian organizational leaders. In W. Mobley & E. Weldon (Eds.), *Advances in Global Leadership* (Vol. 4, pp. 283–297). Emerald Group.

Yukl, G. (2013). *Leadership in organizations*. Pearson.

Yurdasever, E., & Fidan, Y. (2020). KOMB (VUCA) world and new leadership skills. *Journal of the Human and Social Sciences Research*, *9*(2), 1638–1664.

Dr. Hatice Necla Keleş

Agile Leadership in Schools

Abstract
The basic assumption of today's world is change. In this age of rapid change, organizations should strive to catch the agenda and developing their structures. Being ready for change and leading the change are vital conditions for the continuity of today's organizations. The speed of change in economic, social, cultural and technology draws attention to the importance of leadership in organizations. Leadership plays an important role in consciously initiating and accelerating this paradigm shift in organizations and helping to create a future-ready structure. Schools as an educational organization—like organizations in every field—are experiencing a rapid change like never before in history. A key challenge for a school leader in a rapidly changing world is the ability to analyze both the whole and its individual parts of the school and integrate this information to draw consistent conclusions. For this reason, an effective leader needs to catch the speed of change both in the thinking and decision-making processes, which points to the concept of agility. Being agile is important for leadership in this environment of change. Considering the effects of the rapid change experienced on the school's stakeholders and the education process, agile leadership in schools is an important issue in the literature.

Keywords: Agility, leadership, agile leadership, educational leader

> *"Large organizations cannot be versatile. A large organization is effective through its mass rather than through its agility. Fleas can jump many times their own height, but not an elephant."*
>
> *—Peter Drucker*

In B.C. 513 Heraclitus said, "Nothing is permanent except change." Organizations have to be ready to change not only to improve themselves but also to survive in the competitive business environment today. Rapid technological changes, the globalized economy, and changing markets, the rise of e-commerce and e-learning created more opportunities for leaders as well as some threats. Organizations have to answer the questions of what conditions of the current situation are, who must be trusted for accurate information, how the stakeholders are affected, and what kind of path must be followed without wasting time, especially during the COVID-19 pandemic. As a Chinese proverb says, we are passing through a period in which "the mistakes made in the morning must be corrected in the evening."

Emotions of change affect people with their uncertainty and inconsistency. For this reason, managing change alone is not sufficient. The present world almost makes it necessary to manage the unknown (Furr & Dyer, 2014). The only way to deal with competition in a sustainable way is to be agile in today's markets. Agile management is a large global movement that transforms today's world, and what makes this transformation valuable is that it becomes the regular dynamic of the organization and is permanent. Leadership plays important role in consciously initiating and accelerating this paradigm shift in organizations and in creating a structure ready for the future. A successful change also necessitates leadership. To be agile is a broad global movement that transforms working life today, and it is its continuity that makes this transformation valuable.

Leadership

Along with the rapid growth of the high-tech industry, the end of the Cold War marked the beginning of the era of digitalization and globalization. It can be considered that the Cold War accelerated the increasing need for technological solutions with the growth of high-tech industries. Whatever the reasons, information never spread as quickly in the past as it does today. As a result of digitalization and the internet, online shopping, bills are paid and the need to leave the house is reduced, and the information that was scarcely obtained by previous generations is at our fingertips now with one single touch on the screen. This necessitates leaders to be more active and faster in dealing with uncertainty than in the past (Gandhi, 2017).

As the old saying says, "You can take a horse to water, but you cannot force it to drink" is good advice, but it is missing at some points. Better said, "You can lead a horse to water, but you have to lead it to drink." Leading a horse to water is a control problem that necessitates direction and making a horse drink water is a behavioral issue that necessitates leadership (Bawany, 2016). Because, as Niccolo Machiavelli said in his work, Prince (sixteenth century) "The most difficult task to manage, the most dangerous to direct, and the most uncertain success is to lead while introducing innovations." Organizations have changed their focus and methods in leadership development in today's world where the way of doing business is changing rapidly (Kunnath, 2020). Managers must be quick and flexible when faced with unexpected situations. They must also react quickly to rapid changes such as changing rules and regulations, unexpected movements of competitors, changing products, processes, and procedures to cover the needs of the market, cultural differences in employee behavior, and finding and retaining qualified employees as agile leaders (Fernández, 2005).

Agility

In today's economic and technological uncertainty environment, speed is of strategic importance in planning and decision-making processes. In this context, it is necessary to be fast in adapting to new technologies, identifying and meeting stakeholder needs, production, communication and teamwork in all organizations (Akkaya & Bayram, 2021). Agility is the ability of organizations to see unpredictable changes in their environment and adapt their strategies accordingly (Ganguly et al., 2009; Sherehiy et al., 2007). Although different definitions have been made for agility over time, the common expression can be stated as perceiving the unforeseen changes that may be encountered in an efficient and effective manner and acting accordingly.

Agility enables organizations to cope with constant changes, allowing them to thrive in an increasingly volatile, unpredictable, complex, and uncertain world. When managers think about the future and make decisions about it, focusing on the events that are more likely to happen and ignoring the events that are less likely to happen can cause strategic failures. Local and global factors, which are not taken into account in today's world, need to be included in current management paradigms. Traditional paradigms, strategy, and decision-making that are built on assumptions of security and stability need to be updated. To embrace agility is the only way to deal with the competition sustainably in today's markets. Businesses must become as agile as the rapidly changing context in which they find themselves (Denning, 2016). For example, the Spanish clothing brand Zara has proven to be a strong agile organization by filling all of its stores and windows in West America with black creations just 14 days after the September 11 attacks (Akdag, 2020).

The agility definitions in the literature are as follows; agility is about organizations' ability to survive unexpected changes and use changes to their advantage (Sharifi & Zhang, 1999); agility is a skill that enables an organization to survive in an ever-changing and unpredictable environment (Dove, 2001); agility is the ability of an organization to develop and use its internal and inter-agency capabilities (Hooper et al., 2001); agility is an organization's readiness to accept changes in the environment quickly or already, proactively or reactively, through quality, simple and economical components (Conboy & Fitzgerald, 2004). Although the definitions are different, the common expressions they involve can be listed as follows (Meenakshi & Shah, 2018):

(i) Agility is what organizations need to develop, feeling the environment and reacting to change.

(ii) Agility is an ongoing and uninterrupted process.
(iii) Agility is necessary to keep up with the rapidly changing business world uncertainly and unpredictably.
(iv) The reasons leading to agility can be internal or external to the organization.

Agile Leadership

The dynamism of today's world requires being agile in all organizations. Supporting organizational change necessitates flexible and agile leaders increase productivity by keeping up with the pace of change (Boyer & Robert, 2006). How quickly and effectively a business can respond to unforeseen changes in the environment is becoming increasingly important to the long-term success of organizations. Most previous studies that deal with leadership structures in rapidly changing and uncertain environments emphasize agile leadership (Burke, 2018; Denning, 2016; Hayward, 2018; Homey et al., 2010; Lawler & Worley, 2010; McCann et al., 2009; Meenakshi & Shah, 2018; Parker et al., 2015). The realization of agility in organizations appears possible only with the presence of agile leaders (Joiner & Josephs, 2007).

Joiner and Josephs (2007) presented agile leadership as a conceptual model in their study "Leadership Agility-Five Levels of Mastery for Anticipating and Initiating Change." Agile leadership was defined as "the ability to take effective action under complex and rapidly changing conditions." Considering the pace of change, it can be argued that agile leadership has an undeniable importance for corporate sustainability (Wan & Tan, 2021).

Agile leadership is defined as a management style based on teamwork, learning together and giving quick feedback. In addition, focusing on long-term goals such as internal culture, belonging, and feedback, determining new management strategies for changing and uncertain situations, and having the ability to see from different perspectives are the characteristics of an agile leader (Breakspear, 2017). The agile leadership definitions in the literature are shown in Tab. 1.

Tab. 1. Agile Leadership Concept

Agile Leadership	Source
An agile leader is a leader who anticipates and adapts to change, acts quickly and flexibly in a result-oriented manner, and is prone to group and team work.	Higgs & Rowland, 2001

Agile Leadership	Source
The agile leader is quick to respond to changes and developments encountered in the organization, adaptable, reliable, cooperating with and trusting employees, is the leader who transforms it into performance.	Collyer & Warren, 2009
Agile leader takes into account unpredictable environmental conditions and adapts to changes providing leader.	Parker et al., 2015; Tessem, 2014
The agile leader is such that he is self-confident in decision-making, able to draw a roadmap by proactive needs analysis, and individuals with technology skills.	Hayward, 2018
The agile leader is a person who sets guiding principles, develops strategies and have the ability to create mechanisms that will enable a smooth transition to agility is the leader.	Attar & Abdul Kareem, 2020

It was argued that agile leadership is a key factor for corporate success (Parker et al., 2015) and agile leaders affect the global competitiveness of organizations and their future strategic growth plans (Caligiuri, 2013). In their study, Wooten and James (2008) investigated the responses of businesses to the crises in the business environment between 2000 and 2006 and emphasized the importance of agile leadership.

Agile leaders focus on the needs of others and accept other people's perspectives, provide them with the attention they need to achieve their work and personal goals, involve them in decisions (Cooper, 2012). A leadership that focuses on working and learning together as a team, receiving rapid feedback and continuous learning can be explained by agile leadership characteristics. (Breakspear, 2017). The agile leader must be able to remain calm in any situation in order to calm his stressed subordinates. This attitude will help the agile leader to make the right decision in any situation (Prasongko & Tri Adianto, 2019). Agile leadership positively affects flexibility, adaptability, innovation and creativity (Aftab et al., 2022).

Three new skill sets stand out when leading agile organizations. The first is to adopt new personal mindsets and behaviors. Second, providing a suitable environment for teams to work in new ways, and third, incorporating agility into the corporate culture. (Smet et.al, 2018).

There are several benefits associated with the implementation of agile leadership as an operational standard. These include the following (Behrens, 2022):

- Organizations with agile leaders and, by extension, an agile team, are more readily adaptive to change and capable of overcoming obstacles together.
- Everyone is encouraged to use their voice, share insights and roll up their sleeves, so to speak. In that sense, an agile transformation and growth mindset can help team members become agile leaders by way of them operating in a more self-managerial manner.
- Fellow team members are more motivated, actively supported and given the chance to get involved, and inspired to make a meaningful impact on everyday operations.
- Improved communications efficiency, time management, and responsiveness that addresses problems head-on.
- Improved self-organization and, with an agile leader sharing their insights, the adoption of more efficient organizational best practices across teams and/or departments.
- Greater autonomy and self-driven problem-solving rather than waiting for/anticipating others to resolve issues on an individual's behalf.
- The fostering of an organizational culture that adheres to agile principles, thereby operating in a more creative, value-driven manner through empowering its employees.

Changes and innovations are always in the focus of agile leaders. Therefore, during a change, the agile leader will recognize this ahead of time and will not allow the organization to tumble in the change stream (Prasongko & Tri Adianto, 2019). So agile leadership characteristics can be listed as being result-oriented, team-oriented, change-oriented, competent, flexible and fast (Akkaya, 2020) calm, adaptive, innovative, always learning from experience, giving feedback, and not hesitating to help others (Prasongko & Tri Adianto, 2019).

Nine Principles of Agile Leadership

Agile leadership covers a process and is not an approach that can be applied trough learning in a short time. Agile leadership is considered an adaptation process and includes the understanding of development. Nine principles form the basis of agile leadership by Lawrence and Clarke (2019) are explained in Fig. 1.

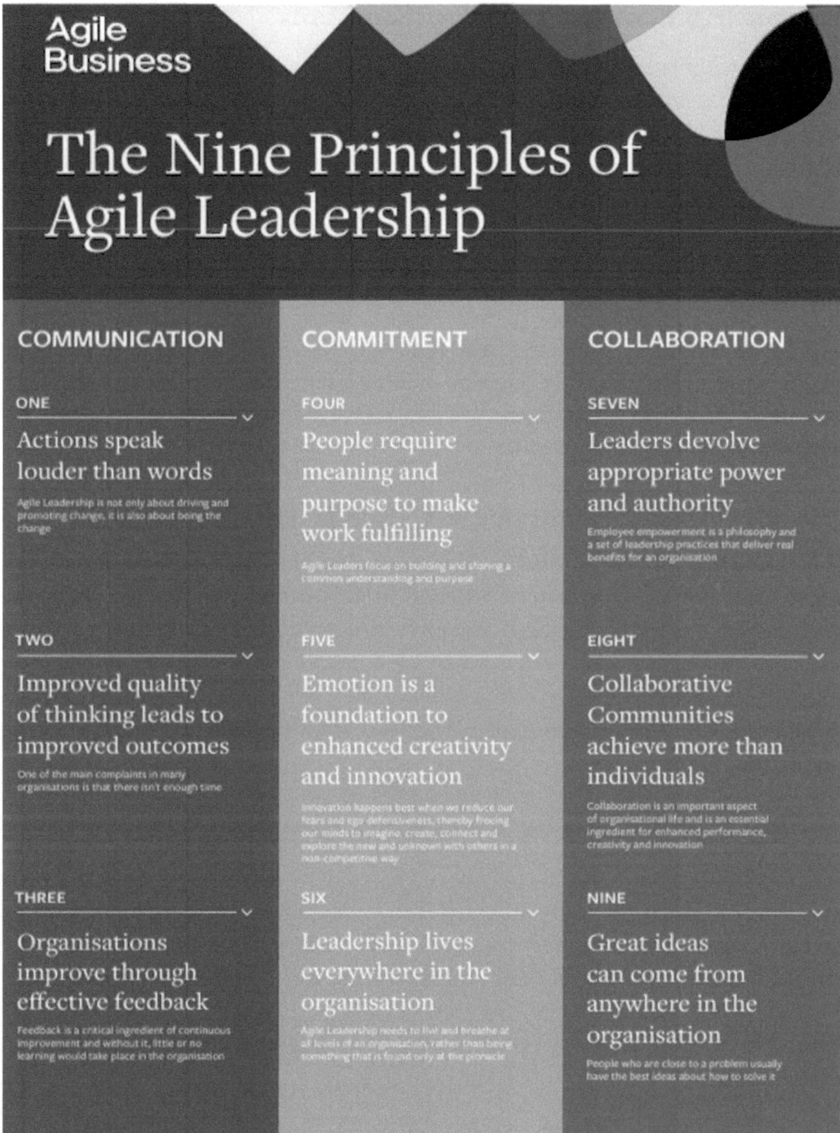

Fig. 1. The Nine Principles of Agile Leadership

Source: https://www.agilebusiness.org/resource/the-nine-principles-of-agile-leadership.html

Principle 1: Actions Speak Louder Than Words

Agile leadership is about initiating, directing and promoting change. Agile leaders are compassionate, courteous, and empathetic to their teams. Agile leaders show interest in and value the suggestions and ideas of employees, constantly improving and renewing themselves by making use of different perspectives (Franz, 2020).

Principle 2: Improved Quality of Thinking Leads to Improved Outcomes

Agile leaders evaluate problems from many aspects. Agile leader does not only rely on digital information in decision-making processes, but also receives data from those closest to the problem without straying from reality. This means reflection and focusing on priorities.

Principle 3: Organizations Improve Through Effective Feedback

Agile leaders make time for the feedback loop and give open, honest, respectful, and effective feedback, which is essential to support and encourage employee development. Giving feedback is essential to support and encourage employee development. Agile leader prefers to create a collaborative network rather than hierarchy. It takes the opinion of the staff on every issue it deems appropriate and exchanges ideas. Agile leader makes decisions with his team and empowers the team (Zink, 2020).

Principle 4: People Require Meaning and Purpose to Make Work Fulfilling

Agile leaders focus on unity of purpose. A meaningful and viable vision of change is important in this respect. Agile leaders must also act by knowing what is in the hearts and minds of their employees and turn it into inspiring action.

Principle 5: Emotion Is a Foundation to Enhanced Creativity and Innovation

Agile leaders inspire everyone on the team to do their best. In addition, knows that emotions are an important part of human experience, and he is aware that adding emotions to the work of individuals increases their potential.

Principle 6: Leadership Lives Everywhere in the Organization

Leadership must permeate all aspects of the organization. Agility necessitates being open to change, innovation, and feedback agile leaders accept the change and, in this direction, encourage the employees to accept the change carefully, following new developments such as science, technology, and professional mobility, adapting them to the organization without losing time (Hanson, 2020; Sarıçam, 2017).

Principle 7: Leaders Devolve Appropriate Power and Authority

Agile leaders know that people are motivated when they are enabled, engaged, and empowered. Empowering individuals, balancing the organization's needs and tensions is an agile leadership skill. Agile leaders realize that empowerment is not all or nothing. They lead as the change requires.

Principle 8: Collaborative Communities Achieve More Than Individuals

Teams of agile leaders are based on trust, respect, and meaningful working relationships. The leader creates an autonomous working environment by providing the requirements to the team.

Principle 9: Great Ideas Can Come from Anywhere in the Organization

Ideas that can be creative, innovative and beneficial to the organization can come from all levels of the organization. People close to the problem in any subject may have the best ideas for a solution. Agile leaders listen to everyone on the team for improvement.

Agile Leadership in Schools

Agile organizations are designed to be proactive and make quick decisions according to internal and external environmental changes to get (Maskell, 2001; Mehdibeigi et al., 2016). Supporting organizational change, increasing productivity and adapting to change can be possible with flexible and agile leaders (Boyer & Robert, 2006).

Agile leadership is a form of management that provides employees with autonomy, aiming to achieve targets and desired outcomes, instead of requiring employees to say anything about doing their job, allowing employees to discover the best way to succeed. With the increased use of mobile phones and the internet, it has become necessary to be able to respond quickly and receive fast feedback. For this reason, agile leadership fits very well with today's conditions (Breakspear, 2017). Agile working approaches that lead to change are required for the complex and deep human dynamics of effective school change. Being agile leaders have critical importance for being able to evaluate the opportunities that emerge, adapt to changing conditions, work in learning and development cycles, and achieve the targets of being renewed (Breakspear, 2016). Global mobility emphasizes that leaders must be agile in order to support an organization's strategies, create value, and adapt quickly to changing organizational needs (Dickmann, 2018).

School leaders are expected to lead development, innovation, and change effectively. Schools are asked to increase students' achievement in literacy and numeracy, develop high-level abilities, cover the needs of an ever-increasing diversity of students, embrace research-based practices, design innovative learning environments, create collaborative environments, and leverage rich data. There was no such an exciting time to be involved in educational change. Investigation of how to cover these demands, which are frequently encountered in this constantly changing context, will require agile leaders to learn with their ability to improve and change learning and teaching in the complex relational environments of contemporary schools (Breakspear, 2016). Considering the school management process, the leadership behaviors of the school administrator are important in conflict management in unpredictable situations (Doeze Jager-van Vliet, 2017).

Agile leaders can adapt quickly by staying calm in change and crisis environments also have the ability to inspire and motivate the people they lead (Prasongko & Tri Adianto, 2019). The problems that organizations face in times of crisis draw attention to the importance of agility in leadership (Aftab et al., 2022). School leaders are expected to lead development, innovation and change effectively. Schools are expected to increase students' achievement in literacy and numeracy, develop high-level abilities, meet the needs of an ever-increasing diversity of students, embrace research-based practices, design innovative learning environments, create collaborative environments, and leverage rich data. Focusing on how to meet constantly changing expectations will need agile leaders to learn with their ability to improve and change learning and teaching in the complex relational environments of contemporary schools (Breakspear, 2016).

An agile leader plays an important role in the organizational culture, organizational climate, managerial processes and the performance of the schools (Hinkler et al., 2011). In order to contribute to agile transformations, "development teams" consisting of teachers can be established apart from the school administrator. In this way, teachers' self-learning and supervision aspects can be strengthened. Considering the current digital age, school administrators' use of the twenty-first century technology equipping them according to their skills can be prioritized. An agile culture in these schools can contribute to its formation. And finally ensuring agile transformations in classroom management and learning for students can help strengthen agility. The support of school administrators to teachers will be important at this point (Özdemir, 2020).

Conclusion

The advancement in technology today causes organizations to gain competitive advantage as long as they act fast, otherwise they lose their market shares (Akkaya et al., 2021). For this reason, agility is becoming increasingly important in educational organizations as in every field. Today, the speed of economy and technology requires all organizations to be fast in terms of adapting to new technologies, meeting stakeholder needs, communication and teamwork. This requires agile leaders in every field. An agile leader is a fast leader who has the speed to learn about changes, teach this knowledge to his teams, and use this knowledge in organizations (Akkaya & Bayram, 2021). Organizations need to create technological agility by following process, information and communication technologies in order to increase their agility in innovation and competitive environments. Working on a team basis and being able to make and implement flexible and fast decisions have made it mandatory for organizations to keep their agility capabilities at the highest level (Akkaya & Bayram, 2021).

Agile leaders are quick, responsive, decision-makers, risk takers, able to handle crises and deal with everything that happens around them (Prasongko & Tri Adianto, 2019). And they sharpen the focus by speeding up the actions of a group of people or organization (Behrens, 2022). Agility is not an all-or-nothing characteristic but must be considered in a continuous aspect. In this respect, agile leadership appears before us as a phenomenon that changes and develops over time. Whether an organization can adapt to future challenges and opportunities depends on the agility level of its leader (Chen et al., 2022). Because it is increasingly important to train agile leaders in organizations and position them to be successful in today's turbulent business settings (Lawrence, 2013).

In today's dynamic environment, agility is needed more than ever. As in every field, the long-term success of educational organizations is related to their ability to adapt quickly and effectively to change. Agility means learning from failures and continuing the fight. This means being ready for success (Cicerali, 2019). As mentioned in the introduction to the chapter, to be agile is a broad global movement that transforms working life today, and it is its continuity that makes this transformation valuable for educational leaders.

References

Aftab, S., Khalid, K., Waheed, A., Aftab, A., & Adnan, A. (2022). Role of agile leadership in managing inter-role conflicts for a satisfying job and life during COVID-19 in a VUCA world. *Frontiers in Psychology, Organizational Psychology, 13*, 1–17. https://doi.org/10.3389/fpsyg.2022.979792

Akdag, A. (2020). *How to be agile brave agile organization practices.* Seçkin Publishing.

Akkaya, B. (2020). Review of leadership styles in perspective of dynamic capabilities: An empirical research on managers in manufacturing firms. *Yönetim Bilimleri Dergisi/Journal of Administrative Sciences, 18*(36), 389–407.

Akkaya, B., & Bayram, V. (2021). Yönetim araştırmalarında postmodern bir liderlik: Çevik liderlik. *Yönetim, Ekonomi ve Pazarlama Araştırmaları Dergisi, 5*(3), 173–188.

Attar, M., & Abdul-Kareem, A. (2020). *Comprehensive strategic management.* Emerald Publishing.

Bawany, S. (2016). Leading in a VUCA business environment. *Leadership Excellence Essentials, 7*, 39–40.

Bessant, J., Knowles, D., Francis, D., & Meredith, S. (2001). *Developing the agile enterprise, agile Manufacturing: The 21st century competitive strategy* (pp. 113–130). Elsevier Science.

Behrens, P. (2022). What is agile leadership? Retrieved from https://www.agileleadershipjourney.com/blog/what-is-agile-leadership

Boyer, M., & Robert, J. (2006). Organizational inertia and dynamic incentives. *Journal of Economic Behavior & Organization, 59*(3), 324–348.

Breakspear, S. (2016). Embracing on agile mindset. *Principal Connections, 20*(1), 30–31.

Breakspear, S. (2017). Embracing agile leadership for learning-how leaders can create impact despite growing complexity. *Australian Educational Leader, 39*(3), 68.

Burke, W. W. (2018). *Organization change: Theory and practice* (5th ed.). Sage.

Bustelo, D., & Vázquez, L. (2006). Agile manufacturing: Industrial case studies in Spain. *Technovation, 26*, 1147–1161.

Caligiuri, P. (2013). Developing culturally agile global business leaders. *Organizational Dynamics, 42*. https://doi.org/10.1016/j.orgdyn.2013.06.002

Chen, X. H., Tee, K., & Chang, V. (2022). Accelerating innovation efficiency through agile leadership: The CEO network effects in China. *Technological Forecasting & Social Change, 179*. https://doi.org/10.1016/j.techfore.2022.121602

Cicerali, E. E. (2019). Çevikliği destekleyen örgütsel kültür özellikleri. *OPUS–Uluslararası Toplum Araştırmaları Dergisi, 11*(18), 2422–2432.

Collyer, S., & Warren, C. (2009). Project management approaches for dynamic environments. *International Journal of Project Management, 27*(4), 355–364.

Cooper, J. D. (2012). *Leadership for follower commitment*. Butterworth-Heinemann.

De Smet, A., Lurie, M., & St George, A. (2018). *Leading agile transformation: The new capabilities leaders need to build 21st-century organizations* (pp. 1–27). McKinsey & Company.

Denning, S., (2016). *Explaining agile senior contributor leadership strategy EDT*. https://www.forbes.com/sites/stevedenning/2016/09/08/explaining-agile/#79faee18301b

Dickmann, M. (2018). Global mobility of the future: Smart, agile, flawless and efficient, London. The RES Forum Annual Report 2018. https://theresforum.com/wp-content/uploads/2018/06/Exec-Summary-RES-Report-2018-RES-FINAL-SINGLE-ONLINE.pdf

Doeze Jager-van Vliet, S. (2017). Proactive and adaptive agility among employees: The relationship with personal and situational factors. *Educational Research, 72*(4), 577–625

Fernández, J. A. (2005). The agile leader: Conditions for succeeding in China. In W. H. Mobley & E. Weldon (Eds.). *Advances in Global Leadership* (pp. 255–275). Emerald Group. https://doi.org/10.1016/S1535-1203(06)04017-2

Furr, N., & Dyer, J. H. (2014). Leading your team into the unknown. *Harvard Business Review, 92*(12), 18.

Gandhi, L. (2017). Human resource challenges in VUCA and SMAC business environment. *ASBM Journal of Management, 10*(1), 1.

Goranson, H. T. (1999). *The agile virtual enterprise: Cases, metrics, tools*. Greenwood.

Hanson, W. (2020, June 17). 15 key qualities that define an "agile" leader. *Forbes*.

Hayward, S. (2018). *The agile leadership. How to Create an agile business in the digital age*. Kogan Page Limited.

Higgs, M., & Rowland, D. (2001). Developing change leaders: Assessing the impact of a development programme. *Journal of Change Management, 2*(1), 47–64.

Hinkler, D., Parente, R., Marion, T. J., & Friar, J. H. (2011). The Role of technology agility on business processes and organizational agilities. In Technology management conference (ITMC), 2011 IEEE international (pp. 67–75). IEEE.

Horney N., Passmore B., & O'Shea, T. (2010). Leadership agility: A business imperative for a VUCA world. *HR People & Strategy, 33*(4), 34–42.

Joiner, B., & Josephs, S. (2007). *Leadership agility-five levels of mastery for anticipating and initiating change*. Wiley.

Kidd, P. T. (1995). *Agile manufacturing: Forging new frontiers*. Addison-Wesley.

Lawler, E. E., & Worley, C. G. (2010). Agility and organization design: A diagnostic framework. *Organizational Dynamics, 39*(2), 194–204.

Lawrence, K. (2013). Developing leaders in a VUCA environment. *UNC Executive Development, 2013*, 1–15.

Lawrence, W., & Clarke, S. (2019). Leveraging agile leadership for digital transformation. In *Conference: 10th Annual MSBM Roundtable CEO*. https://doi.org/10.13140/RG.2.2.35440.92162.

Lin, C. T., Chiu, H., & Tseng, Y. H. (2006). Agility evaluation using fuzzy logic. *International Journal of Production Economics, 101*(2), 353–368.

Maskell, B. (2001). The age of agile manufacturing. *Supply Chain Management, 6*(1), 5–11.

McCann, J., Selsky, J., & Lee, J. (2009). Building agility, resilience and performance in turbulent environments. *HR People & Strategy, 32*(2), 45–51.

Mehdibeigi, N., Dehghani, M., & Yaghoubi, N. M. (2016). Customer knowledge management and organization's effectiveness: Explaining the mediator role of organizational agility. *Procedia Social and Behavioral Sciences, 230*, 94–103.

Jakhar, M., & Bharadwaj, S. S. (2018). Agile capabilities for innovative and imitating firms: A framework for mobile handset industry. *Journal of Management Research, 18*(2), 114–126.

Özdemir, A. N. (2020). *Okul yöneticilerinin çevik liderlik özelliklerinin örgütsel bağlılığa etkisi: İngiltere ve Türkiye okullarında karşılaştırmalı bir analiz* [Unpublished PhD Thesis]. Marmara Üniversitesi, Eğitim Bilimleri Enstitüsü, İstanbul.

Parker, D. W., Holesgrove, M., & Pathak, R. (2015). Improving productivity with self-organised teams and agile leadership. *International Journal of Productivity and Performance Management, 64*(1).

Prasongko, A., & Adianto, T. (2019). The role of the agile leadership model as a competitive advantage for the future leader in the era of globalization and industrial revolution 4.0. *Jurnal Pertahanan: Media Informasi ttg Kajian &*

Strategi Pertahanan yang Mengedepankan Identity, Nasionalism & Integrity, *5*(3), 126–133.

Sarıçam, E. (2017). *Lider yönetici*. AZ Yayıncılık.

Sharifi, H., & Zhang, Z. (1999). A methodology for achieving agility in manufacturing organisations: An introduction. *International Journal of Production Economics*, *62*(1), 7–22.

Tessem, B. (2014). Individual empowerment of agile and non-agile software developers in small teams. *Information and Software Technology*, *56*(8), 873–889.

Wan, H. L., & Tan, N. N. (2021). Agile leadership and bootlegging behavior: Does leadership coping dynamics matter? In N. Ferreira, I. L. Potgieter & M. Coetzee (Eds.), *Agile coping in the digital workplace* (pp. 187–202). Springer.

Wooten, L. P., & James, E. H. (2008). Linking crisis management and leadership competencies: The role of human resource development. *Advances in Developing Human Resources*, *10*(3), 352–379. https://doi.org/10.1177/1523422308316450

Yusuf, Y. Y., Sahardi, M., & Gunasekaran, A. (1999). Agile manufacturing: The drivers, concepts and attributes. *International Journal of Production Economics*, *62*, 33–43.

Zink, P. (2020, June 17). 15 key qualities that define an "agile" leader. *Forbes*.

Dr. B. Çağla Garipağaoğlu

Leading Educational Institutions in the Age of Chaos & Complexity with Quantum Leadership

Abstract
Alan Watts argues, "A completely predictable future is already the past." However, according to Albert Einstein, although the future is unknown and uncertain, it is still somewhat predictable. He argues that all we need to do is to look back upon the past where the seeds of the future were planted; try to comprehend the "big picture" and to notice the revealing patterns. Parents, students and communities are increasingly demanding their schools to exhibit high reliability at the face of an adversity and to address not only the crises that are already expected, but also to address those that cannot be foreseen (McManus et al., 2008). So, eventually, like any other organization, educational institutions understood that they can only rely on quantum leadership to successfully navigate through uncertain, volatile and very demanding environment that they have to operating in. Quantum leadership is a type of leadership guided by the defining principles of quantum physics, chaos, and complexity. This chapter is, therefore, dedicated to explain the fundamentals of quantum leadership and characteristics of quantum leaders who inspire others to thrive during periods of great change and uncertainty by drawing upon the theory of chaos and complexity.

Keywords: Quantum age, chaos, complexity, educational leadership

The Age of Quantum

Without clearly recognizing it, educational leaders are challenged by rapid change, uncertainty, intense interconnectivity, greatest technological upheaval, de-centralization, globalization and greater demands for ethics and meaning. This compels them to adopt a brand-new leadership paradigm that does not only require a new thinking but also new metaphors, new assumptions and new values that is more adept to deal with those forces creatively (Uhl-Bien et al., 2007). Both for educational leaders and students, there is no better model for such a paradigm shift than the revolution in scientific thinking that gave us quantum physics, chaos, and complexity science (Kuhn, 1997).

Since the time of the Greeks, science attempted to understand the universe by breaking it down into its smallest and the most fundamental parts with the belief that, at a fundamental level, the system was ordered, linear and predictable (Nagai, 2021). For centuries, driven by this idea of reductionism, most prominent scientists like Galileo, Descartes, Newton and Laplace claimed that the universe was characterized by mechanistic determinism and the future was just the extension of the past (Albright, 2013). Laplace claimed that a demon with the ability to recognize all the initial conditions and restraints could have predicted the future with 100% certainty (Feinberg et al., 1992). Therefore, according to Laplace, for a man with such a demonic ability, nothing was uncertain and "freedom of will" was nothing but an illusion (Van Cleave, 2019). This paradigm had a strong influence on setting the course of scientific thinking for years to come. And by the early nineteenth century, determinism had become firmly entrenched among many scientists (Uhl-Bien et al., 2007). Our attempts to make the world comprehensible caused us to assume that the world is regular, simple, periodic, and predictable (Gimbel, 2020). Despite the fact that this way of thinking is still valid in some cases, universal application of it brings about philosophical gibberish like Laplace's demon (Nagai, 2021). Hence, for more than 100 years, the Newtonian-inspired way of reductionist understanding of the world dominated the world's main approach to schooling and education.

Chaos Theory

Chaos theory argued that even for Newton's deterministic systems, the events are unforeseen, and the future is still unpredictable. Yet, the problem of unpredictability is not just about complexity, it is about the instability and randomness of the living dynamic systems. Indeed, scientists regularly tackle complex problems with ease, so the problem, according to Nagai (2021), must be something much more fundamental than the complexity. The truth is that the universe is not only complex but also chaotic and this is exactly why the future cannot be predicted by the extrapolation of the past (Nagai, 2021). In layman's terms, the word "chaos" means mere disorder but among the scientists of complexity, it is a technical term indicating randomness and instability in the systems that behave in an unpredictable manner according to "deterministic" non-linear rules (Gleick, 2011). To make sense of chaos, consider a game of sports in which the players play according to specific rules but just because they behave chaotically, nobody can guess which team will win over the other with 100% certainty. Even in the game where it is evident that a certain team is clearly superior over the other, a little mistake or unexpected move might be amplified to result in a surprising defeat of the favorite team. Indeed, it is why most people are charmed by games of sports. Despite all the determinate rules, because of their chaotic nature, the end results are never completely predictable in a game of sports. In short, as in the case of games of sports, chaos theory fundamentally proves us that there are some certain restrains to our knowledge and our ability to prediction (Boeing, 2015). Think about the emergence of MOOCs (Massive Open Online Courses) and micro-schools! Who could have thought the classical brick-and-mortar education system would transform gradually into MOOCs and the pandemic would be responsible for this drastic change? Or who could have guessed micro-schools would become the new and powerful players of the new education system? Or ChatGPT would compel all the educators around the globe to formulate different assessment methods for their students. All these developments were not even on the radar of most educators and educational leaders. Because of the chaotic nature of the systems, we create, and we live in, it is simply near-impossible for us to see how the events will unfold for us in future.

Another feature of chaos is the sensitive dependence on initial conditions (Lorenz, 1963). This is called "butterfly effect" and means that a flap of a butterfly's wings might generate a tornado in a distant place on earth (Lorenz, 1972). Chaos theory explains the behavior of dynamic systems like weather which are extremely sensitive to initial conditions. The weather is a chaotic system. A tiniest fluctuation (a butterfly effect) in air may have tremendous effects in a completely distant part of the world. This explains why scientists find it so difficult to predict the weather and it is why a weather forecaster can scarcely predict weather five weeks ahead while they can make better predictions for a day later. To predict future situations, they need to know the initial situation in its finest detail. This is actually why Lorenz (1972) argues that while the "present" determines the future, the "approximate present" does not. Even though the systems are deterministic, a small variation in the initial condition may result in widely different outcomes in the system by making the prediction of future behavior and outcomes nearly impossible. Moreover, as time increases, uncertainty in predictions increase exponentially leading predictions to become more random over time (Dunstone, 2019). Any missing detail in the present, no matter how small it is, can have drastic effects on the future predictions, hence making forecasts typically relatively short range and unreliable for any dynamic system—biological, economic, or social (Dunstone, 2019). It must be remembered that, as Gaiman and Pratchett (2014) wrote in "Good Omens," this is why the events that changed the world were not the vast atomic bombs, lunatic politicians, huge earthquakes, or the mass population movements. In fact, the things that change the world eventually, according to the chaos theory, were the very tiny little details that have never been thought crucial by people who held an old-fashioned Newtonian view which is totally out of touch with the quantum physics (Gaiman & Pratchett, 2014).

Chaos theory has also shaken science to its foundations with the realization that very simple rules can trigger extraordinary intricate behavior: river and fractals (Waldrop, 1993). Hidden deep within the chaos, there are surprising recurring patterns revealing itself in time. Patterns that, if we are ever able to fully understand them (like Laplace's demon), might lead to some deeper revelations. Self-organizing, fractal, non-linear, constant feedback loops, and heavy reliance on the initial situations are inherently unpredictable, uncontrollable but still understandable only in the hindsight and from the most general way (Waldrop, 1993). In order to discover novelties and create positive values, we should be able to zoom in or out to different levels of scale and take a look at the events from outside. Because it is only then fractals—self-repeating essential structures/ patterns become visible (Mandelbrot, 1983) and we can make better sense of the world we are living in. Only if we can excel at standing back from the problem/ situation, look for the bigger picture, relate with the wider context and reframe the problem (Zohar, 2016), then we can somehow understand how the future will unfold—yet still only in the most general way.

Complexity Theory

During the 1990s, complexity theory grew out of chaos theory (Boeing, 2015). While they both have the similar principles, in the end, they are not the same beasts (Boeing, 2015). Unlike chaotic systems, complex systems retain some trace of their initial conditions and previous states through path dependence (Boeing, 2015). Although they are both unpredictable, they are unpredictable in a different way. Complex systems have the ability to surprise through creativity, novelty and inception (Boeing, 2015). The edge of chaos is the most critical region in the theory of complexity where a small change can either draw the system into a chaotic behavior or lock it into a constant behavior (Koç & Sönmez, 2017). According to Jencks (1995), chaotic and complex systems are more likely to generate creativity than linear systems as the creativity is balanced at the knife-edge between predictability and randomness. This is indeed why the educational leaders must adopt a more holistic view of the system and embrace disruptive innovations (Christensen et al., 2008) if they are to keep up with the continuously changing and demanding requirements of their ecosystem.

Complexity theory examines large open systems made of many interacting parts and states that complexity brings "order for free." Kauffman (1993) has pioneered the idea that complexity gives rise to "self-organization" through the wisdom of "swarm intelligence." He contends that living systems, and the environments interact with each other, hence cause the emergence of spontaneous, self-generated order—"order for free" that plays a major role in evolution, alongside natural selection (Boeing, 2015). Chaotic complex systems introduce us the concept of "destruct in order to construct" (Koç & Sönmez, 2017). Science seeks order in chaos and finds it, but that order is so complex that it seems to trigger a new disorder (Gimbel, 2020). The chaos theory defends that every order tends to be destroyed and cannot be preserved without an external authority or power; if an order maintains its existence, it can be said that there are some authorities resisting and interrupting the tendency of destruction and supporting the order (Koç & Sönmez, 2017). With all that being said, now it makes much more sense for educational leaders to avoid the adding-on approach in education. Despite the valid demand for change in educational system, expecting the system to change merely by adding another layer on top of the existing does not make any sense at all from the quantum perspective. As advocated by Prensky (2014), "add-on" approach in the educational curriculum per se is not what we need to make our education system better and wiser, rather what we need is to adopt a more holistic system view in every single part of the educational system. According to quantum paradigm, that is exactly the reason why the system must first be "destructed" just to be "re-constructed" from the scratch. This is actually why some professionals (e.g., the former chairman and CEO of General Electric (GE) Jack Welch and the former CEO of Intel Andy Grove) fiercely argue that systems should encourage the act of intentional self-cannibalization and forced obsolescence. Harvard Business School's Professor Dr. Christensen (1997) calls this "survival by suicide."

As we all know, our current "academic" educational system—designed originally for the nineteenth century—no longer addresses to our quickly arriving future (Prensky, 2016). It does not produce the people we need; and it is essentially a system training the mere thinkers but not the doers (Prensky, 2016). The fundamental question is that "Are we going to fix it, or replace it with something new?" As argued by Hamel (2001), the future can't be predicted, and the goal of foreseeing the future and preparing for it perfectly are just unrealizable, but it can be envisioned and brought lovingly into being. However, according to Meadows (2008), the idea of making a complex system do just what you want it to do can be achieved only temporarily, at best. We can never fully understand our world, not in the way our reductionist science has led us to expect, but we can surely co-create the future and influence it continuously. We can neither control a system nor impose our will upon it, but we can design and redesign it continuously and eventually (Meadows, 2008). We can listen to what the system tells us and discover how its properties and our values can work together to bring forth something much better than our imagination or will (Meadows, 2008). So, all we have to do is to embrace the swarm intelligence, surprises and learn and even profit from them. By adopting value propositions of the quantum paradigm that prioritize the system thinking (Senge, 1990), rather than letting our destiny to be controlled by the factors that are out of our own control, we can create an ecosystem that creatively prospers in future.

Mechanical determinism collapses gradually during the twentieth century. The uncertainty principle in quantum mechanics denied the determinism at a microscopic level, and the butterfly effect in the science of complexity denied the determinism at a macroscopic level (Nagai, 2021). Complexity is the science of understanding how independent parts interact with each other to influence each other and the whole (Uhl-Bien et al., 2007). Organisms constantly adapt to each other through evolution, thereby organizing themselves into an exquisitely tuned ecosystem (Waldrop, 1993). Self-organizing systems are adaptive and actively try to turn whatever happens to their advantage. Complex systems have somehow found a way to acquire the ability to bring order and chaos into balance. The edge of chaos is where new ideas and innovations are forever nibbling on the status quo (Waldrop, 1993). The edge of chaos is the constantly shifting battle zone between stagnation and anarchy, the one place where a complex system can be spontaneous, adaptive and alive (Waldrop, 1993).

The Quantum Paradigm

Older management traditions studies are the fragmented pieces of a whole. The main thesis of the traditional Newtonian paradigm is that systems tend toward hierarchy, order, stability (homeostasis), and entropy and the order is maintained by clearly specifying the "boundaries and roles"; and the change happens to come into being only through the redefinition of those "boundaries and roles" (Curtin, 2013). Ye, this classical Newton paradigm and logic that gave rise to the Industrial Revolution—assuming corporations and human beings function like machinery parts—are no longer valid. The quantum paradigm, on the other hand, regards the organization and its members as interconnected energetic beings (Bohm, 1980). These connections are non-linear, non-hierarchical and can be enhanced by collaboration and interaction. This view is supported by quantum physics, chaos and complexity science which suggest a different set of metaphors. The quantum paradigm states that nothing is certain, events cannot be foreseen, control is a delusion, and change is inevitable. Quantum physics and the thinking stress the holistic view and the creative potential of uncertainty in our current complex, volatile and ambiguous world and such breaks away from the Newtonian clockwork machines perspective that emphasizing on fragmented parts. It characterizes the universe as constantly evolving, holistic, interconnected, self-organizing systems that advance with trial and error. Hence, the shift toward Quantum Organization from a Newtonian Organization requires a more dynamic model that embraces uncertainty and countless possibilities. It requires to integrate the most effective traditional leadership practices with recent discoveries in quantum physics, chaos, and complexity science (Porter-O'Grady & Malloch, 2017). A quantum leader is a human-centered person who is aware of the fact that s/he is the "co-creator" of every moment of reality together with others and, as such, is able to co-create and influence the reality and the future continuously and thus, can thrive on uncertainty (Zohar, 2022). They think ahead by formulating many scenarios for what the future might hold, encourage questioning, trial-errors, and focus on meaning, purpose, and motivation (Porter-O'Grady, 1999). According to Zohar (2022), becoming a quantum leader also requires becoming a good person with integrity whom others can trust on, and inspired from.

As times change, educational leaders now face the greatest need for creative restructuring since the Industrial Revolution that requires the need for a quantum leap in their thinking. Quantum organizations are what complexity science calls "complex adaptive systems-CAS" (Weick, 1995). Since the external forces are not in control of any organization, unforeseen and complicated, leaders need to design and redesign their systems and operations and to realign their cultures to address the same. This is why leaders should start off with broad goals, get a general idea (not detailed) about what they want to achieve and plan the details when they need to be planned, directly before implementation. Wasting too much time on rules, details, the order may hinder creativity. Teachers focusing on following the detailed curriculum or lesson plans sent by the headquarters will never be able to bring in their own thoughts to the school processes. This may work in a factory where following specifications is essential, but it doesn't work well if you want your teachers and students to be creative or innovative. Instead, quantum leaders give their teachers and the management team basic parameters to follow. Instead of dictating, they guide and let their teachers/students bring in their creativity and own their learning process.

Traditional leadership paradigms state that cause-and-effect relationship is linear; and an organization's internal dynamic influence members but assumes that only the leaders may have an effect on the internal dynamics. The quantum perspective, on the other hand, emphasizes free-flowing interaction and co-determination; members affect the internal dynamic as much as the internal dynamic affects members (Curtin, 2011). The quantum perspective emphasizes that all organizations consist of a world of connections, interactions and relationships. And the only meaningful change comes from within and occurs first among various members (Curtin, 2011).

Most organizations rely on standard organizational designs that use a top-down hierarchy to control the organization. Quantum Leadership proposes a more flexible and flatter model that enables people and organizations to change as the environment changes. The employees aren't told how to organize. They work with whomever they need to get their work done. They build the organization as they go and adapt it when the need arises. Organizations are surprisingly complex structures that function as living, non-linear, dynamic systems. Therefore, employees no longer are simple obedient agents of a hierarchical organization, they rather are interconnected and interacting participants of more of an invisible, intangible, unforeseen complicated morphogenetic field. Bohm (1980) believed that the general tendency for individuals and social groups to see one another as different and separate entities was a major source of conflict in the Newtonian world. The breaking down of our "old" Newtonian education system and the breaking through of the "new" quantum system requires equality, communion, collaboration, synchronicity, expansiveness, abundance, wholeness, mutuality, intuitive knowing, and love instead of individualism, patriarchy, a scarcity mentality, competition or the dualistic constructs of superior/inferior, win/lose, good/bad, and domination/submission (Curtin, 2013). This shift to quantum paradigm, indeed heralds a hopeful future for our world where people might come to recognize the essential interrelatedness of all things and would join to build a more holistic and harmonious world.

Conclusion

With all that said, it's important to note that all of these do not mean that the quantum paradigm will or should replace the traditional paradigm (Curtin, 2011). Quantum and Newtonian paradigm each addresses different needs of the organizational life and leaders should appreciate the both and become skilled at accommodating both approaches. While the traditional approach is frequently useful for situations that are subject to control, the quantum paradigm is more useful for understanding unfamiliar events that arise during turbulent times in complex adaptive systems where there are strong pressures to change, events seem chaotic, goals are ambiguous, and order seems to emerge of its own accord and in its own time (Kiel & Elliot, 1996). It must be remembered that because our minds are bounded by cognitive and other limitations—"bounded rationality" (Herbert, 1957), our ability to deal with problems of chaotic complex systems are also bounded and this is exactly why we tend to settle for the sub-optimal choices. Perhaps, as artificial intelligence and big data improves, our ability to predict the future or deal with the increasingly complex problems of the future might improve (Dunstone, 2019). However, more often than you might think, simple heuristics may still help us make better, and quicker decisions than theoretical optimal AI procedures (Gigerenzer & Selten, 2002). Hence, despite the marvels of AI and the big data, we should never underestimate the power of heuristics, frequent feedbacks and learn from our mistakes. We should act from our principles, deep beliefs, and live life accordingly (Meadows, 2008). Sometimes, we should risk hitting the ground harder and faster and destruct in order to construct for better rather than expending the much-needed energy in the futile attempt to preserve the existing system and let the system bring its new order "for free" through the wisdom of swarm intelligence.

References

Albright, J. R. (2013). Causality in physics. In A. L. C. Runehov & L. Oviedo (Eds.), *Encyclopedia of sciences and religions*. Springer. https://doi.org/10.1007/978-1-4020-8265-8_719

Boeing, G. (2015). *Chaos theory and the logistic map*. https://geoffboeing.com/2015/03/chaos-theory-logistic-map/

Bohm, D. (1980). *Wholeness and the implicate order*. Routledge & Kegan Paul.

Christensen, C. M. (1997). *The innovator's dilemma: When new technologies cause great firms to fail*. Harvard Business School Press.

Christensen, C. M., Horn, M. B., & Johnson, C. W. (2008). *Disrupting class: How disruptive innovation will change the way the world learns*. McGrow-Hill.

Curtin, L. (2013). Quantum leadership: Upside down. *American Nurse Journal*. https://www.myamericannurse.com/quantum-leadership-upside-down/

Dunstone, J. (2019). *Chaos theory and how to predict the future*. Square Holls. https://squareholes.com/blog/2019/07/05/chaos-and-how-to-predict-the-future/

Feinberg, G., Lavine, S., & Albert, D. (1992). Knowledge of the past and future. *The Journal of Philosophy*, *89*(12), 607–642. https://doi.org/10.2307/2940898

Gaiman, N., & Pratchett, T. (2014). *Good Omens*. Gollancz.

Gigerenzer, G., & Selten, R. (2002). *Bounded rationality*. MIT Press.

Gimbel, S. (2020). *Order in randomness: Fractals and chaotic systems*. WondriumDaily. https://www.wondriumdaily.com/order-in-randomness-fractals-and-chaotic-systems/

Gleick, J. (2011). *Chaos: Making a new science*. Open Road Media.

Hamel, G. (2001). Leading the revolution: An interview with Gary Hamel, *Strategy & Leadership*, *29*(1), 4–10. https://doi.org/10.1108/10878570110367141

Herbert, S. (1957). *Models of man: Social and rational*. Wiley.

Jencks, C. (1995). *The architecture of the jumping universe*. Academy Editions.

Kauffman, S. A. (1993). *The origins of order: Self-organization and selection in evolution*. Oxford University Press.

Kiel, L. D., & Elliott, E. W. (1996). *Chaos theory in the social sciences: Foundations and applications*. University of Michigan Press.

Koç, Ç., & Sönmez, M. (2017). The architect of today on the edge of chaos: An approach through chaos theory. *Gazi University Journal of Science Part B: Art, Humanities, Design and Planning*, *5*(1), 1–7. https://dergipark.org.tr/tr/download/article-file/290252

Kuhn, T. S. (1997). *The structure of scientific revolutions* (3rd ed.). University of Chicago Press.

Lorenz, E. N. (1963). Deterministic nonperiodic flow. *Journal of the Atmospheric Sciences*, *20*, 130–141.

Lorenz, E. N. (1972). Predictability: Does the flap of a butterfly's wings in Brazil set off a tornado in Texas. *American Association for the Advancement of Science*, *6*(13), 1–3. https://static.gymportalen.dk/sites/lru.dk/files/lru/132_kap6_lorenz_artikel_the_butterfly_effect.pdf (gymportalen.dk)

Mandelbrot, B. B. (1983). *The fractal geometry of nature*. Freeman.

McManus, S., Seville, E., Vargo, J., & Brunsdon, D. (2008). A facilitated process for improving organizational resilience. *Natural Hazards Review*, *9*(2), 81–90. https://doi.org/10.1061/(ASCE)1527-6988(2008)9:2(81)

Meadows, D. (2008). *Thinking in systems: A Primer*. Chelsea Green Publishing.

Nagai, T. (2021). *Chaos and determinism*. https://www.nagaitoshiya.com/en/2000/chaos-determinism/

Porter-O'Grady, T. (1999). Quantum leadership. New roles for a new age. *The Journal of Nursing Administration*, *29*(10), 37–42. https://doi.org/10.1097/00005110-199910000-00008

Porter-O'Grady, T., & Malloch, K. (2017). *Quantum leadership*. Jones and Bartlett.

Prensky, M. (2014). The world needs a new curriculum. *Educational Technology*, *54*(4), 3–15.

Prensky, M. (2016). *Education to better their world: Unleashing the power of 21st-century kids*. Teachers College Press.

Senge, P. M. (1990). *The fifth discipline: The art and practice of the learning organization*. Doubleday.

Uhl-Bien, M., Marion, R., & McKelvey, B. (2007). *Complexity leadership theory: Shifting leadership from the Industrial Age to the Knowledge Era*. https://digitalcommons.unl.edu/cgi/viewcontent.cgi?article=1017&context=leadershipfacpub

Van Cleave, M. J. (2019). The problem of free will and determinism. In M. J. Van Cleave, P. M. Jurczak, C. Schneck, & D. Sjoquist (Eds.), *Introduction to philosophy* (pp. 81–108). UCF Pressbooks. https://pressbooks.online.ucf.edu/introductiontophilosophy/chapter/the-problem-of-free-will-and-determinism/

Waldrop, M. M. (1993). *Complexity: The emerging science at the edge of order and chaos*. Simon and Schuster.

Weick, K. E. (1995). *Sensemaking in organizations*. Sage Publications.

Zohar, D. (2016). *The quantum leader*. Prometheus Books.

Zohar, D. (2022). *Zero distance: Management in the quantum age*. Palgrave Macmillan.

Hatice Çakır and Dr. Özge Öztekin Bayır

Building Resilience in Education: Ensuring Quality Learning amid Crises and Disruptions

Abstract

Resilient education systems refer to the ability of education systems facing challenges such as natural disasters, economic crises, and pandemics to maintain the quality of education without negatively affecting students' learning outcomes. These challenges can cause interruptions in education delivery and negatively impact students' learning outcomes. Therefore, it is important to create resilient education systems to ensure that education remains accessible and of high quality for all students. Millions of children worldwide have had their education disrupted because of the COVID-19 pandemic, which has drawn attention to the weaknesses and disparities in national education systems. However, this has also presented us with possibilities to deal with the fundamental concerns of unequal learning opportunities and to construct education systems that are more adaptable and resilient. In this context, a resilient education system and its components have been explained in this chapter, and recommendations have been made for the education system to become more resilient to potential future crises.

Keywords: Resilient education systems, learning outcomes, COVID-19 pandemic, pandemic impact on education

The resilient educational system is defined as a system needed to cope with the inequality of opportunity in education and to offer an equitable education environment to every student in case of a crisis. In more general terms, it is about ensuring that the delivery of education is not interrupted in cases such as pandemics, endemics, natural disasters, seasonally changing weather and climate conditions (extreme hot or cold weather, air pollution), or emergencies (war, internal uprising) (Van Der Graaf et al., 2021). A resilient educational system is based on the concept of *resilience.*

Although the international literature offers no consensus on the concept of resilience, it refers to accomplishment despite various risk factors and involves the capacity to rise again after facing challenges or fear; this characteristic involves the combination of the terms passion and perseverance (Yamashita, 2018). The significance of resilience stems from the need to strengthen the capabilities of people and societies to tolerate and adapt to a wide variety of challenges

(Frankenberger et al., 2014). It is also defined as the ability to withstand, adapt, and recover from the stress created by the risky situations that children, families and teachers face (Bethke et al., 2015). As such, the resilient education system includes dimensions such as delivering educational content to every learner, continuing the educational process without interruption, and supporting families, students, and teachers. Such an education system needs a consistent, quality, and hybrid approach to enable simultaneous face-to-face and distance learning and to monitor student engagement and well-being during quarantines (Dabrowski et al., 2020). Therefore, hybrid and distance education are essential in the resilient education model. Hybrid education combines traditional face-to-face education with digital technology and blends it with distance education, which can also be defined as coeducation or blended education (Hall & Villareal, 2015). On the other hand, distance education provides an alternative learning environment by providing flexibility in learning and teaching processes regardless of time or place. After COVID-19, the concept of "emergency distance education" has become a part of our lives (Bozkurt & Sharma, 2020; Mısırlı & Ergüleç, 2021; Tekel et al., 2022).

One of the most important social institutions to be negatively impacted by the pandemic was the educational system (Karadağ & Yücel; Kırmızıgül, 2020). With the closure of schools and the interruption of education due to the worldwide spread of COVID-19, many countries have begun to question the deficits of their education systems and the inequalities of opportunity these deficits create. The resiliency of education systems worldwide was tested at every stage of education, from preschool to postgraduate study, during the COVID-19 pandemic, which caused a very different and unconventional situation for learning and teaching (Naidu, 2021). Although implementing distance and hybrid education models during the pandemic seems to have created very few access problems for socioeconomically advantaged students, for students with lower socioeconomic status, access has been a major hurdle. Although distance and hybrid education models have been put into practice to solve this problem, according to UNICEF, OECD, and World Bank data, only a third of the developed countries and only a quarter of the developing countries with middle or lower socioeconomic status have been able to implement hybrid education (UNICEF, 2022). These data reveal the importance of building a resilient education system so every student can access emergency education.

Elements of the Resilient Education System

A resilient education system is one that can transform itself in the face of difficulties and adapt to challenging situations (Van Der Graaf et al., 2021). COVID-19 has interrupted education worldwide, leading educators to revise their educational practices, re-think the concept of resilient education, and make alternative plans not to have their delivery of education interrupted. This revision demonstrates that eradicating inequality in educational settings is attainable by developing a resilient education system that can be effectively implemented by combining specific features. Tab. 1 lists these elements and their subcomponents.

Tab. 1. Elements of the Resilient Education System (Tarricone et al., 2021)

Education System	Instruction	Representatives
o Emergency education planning o Collaboration and coordination o Communication o ICT infrastructure o School buildings protocols o Monitoring	o Curriculum o TV, radio and print materials o Digital technology o Hybrid learning o Evaluation & Assessment	o National and local governance o Intergovernmental organizations and NGOs o Schools and school leaders o Teachers o Parents o Children

As shown in Tab. 1, the education system, instruction, and representatives are the key components of a resilient education system. Six steps are mentioned in the USA High-Quality School Emergency Operations Plans Development Guide for planning education system-wide emergency education (Readiness and Emergency Management for Schools (REMS) Technical Assistance (TA) Center, 2019). These steps are:

(i) Building a collaborative planning team
(ii) Understanding the current situation
(iii) Setting goals and objectives
(iv) Developing a plan
(v) Preparation, review and approval of the plan
(vi) Implementation and evaluation of the plan

Along with the steps necessary for the planning of education in emergencies, establishing mutual communication, ensuring cooperation and coordination, the existence of ICT infrastructure and its control, the physical conditions of schools, monitoring, and evaluating studies, the sub-structures deemed necessary for the construction of a resilient education system are some other conducive actions to be taken during this process.

The instructional dimension covers all activities, including how educators convey information to students, how students learn the transferred information, and evaluating these processes (Tarricone et al., 2021). Among these activities, the delivery of educational resources to all students is the primary step. For all children to have access to resources, the curriculum to be created within the resilient education system must be publishable on all platform types. In addition, the curriculum should be delivered to every child by telephone, television, print materials, and digital tools. However, delivering the entire curriculum to children with digital tools is insufficient. Some regions worldwide, including Turkey, lack a reliable internet connection or electricity and may have difficulty accessing printed resources. One of the most important objectives of the resilient education system is reaching every child and trying to eliminate the inequality of opportunity that arises due to economic and geographical reasons. Therefore, the implementation of hybrid education in locations where there is no internet access is also included in the instruction element. The instruction process should not be perceived as a process in which knowledge is uniformly transferred. This process should involve interacting with students. Once the emergency is over, performing a comprehensive assessment as soon as children return to school is an essential step in identifying damage during learning. At the end of the evaluation process, developmental differences in students can be determined, and appropriate early intervention work can be initiated.

Representatives also play an important role in creating a resilient education system. These representatives include national and local governments, international organizations and non-governmental organizations, schools, school administrators, teachers, and parents with decision-making authority throughout the education process. Leaders with decision-making skills have an important place in the resilient education system because a quick switch to online education requires a formidable and reliable structure (Makoe, 2023). Leading such a significant change requires resilient leaders who are "capable of facing difficult situations, learning from and empowering through challenges" (Reed, 2018).

In a resilient education system, the ability of a school to adapt to and prevail over problems, whether those challenges originate from within the school or from the outside world, depends heavily on the school administrator. As a multifaceted, scientifically grounded construct, resilience provides school leaders a potent toolkit for overcoming adversity. In this context, school leaders' resilience capacity relies on three sources: (i) personal values; (ii) personal efficacy; and

(iii) personal energy. These three factors, at any given time, determine the leader's potential to go forward (Patterson & Kelleher, 2005). It is also pointed out that each school principal is an individual with their own experiences, beliefs, and habits that they bring to their position (Sergiovanni, 2001). These differences also create a rich diversity of experience in leadership in the management of educational organizations. In order to build resilient educational organizations and a resilient education system, if we consider it from a broader perspective, resilient leaders are needed. In this context, there are a few alternative ways that principals of schools can pursue in order to encourage resiliency in their schools:

(i) *Building a culture of resilience:* When principals create a culture of continuous improvement in their schools, they can spread a positive attitude and stress how important it is to learn from mistakes. Dweck and Yeager (2019) also state that having a growth mindset relates to increased resilience as well as the ability to adapt to different environments.

(ii) *Fostering relationships and collaboration:* Strong relationships can be built by school leaders with their pupils, staff, parents, and other members of the community. In times of crisis, this can develop a sense of trust and teamwork, which can be crucial factors. Principals who are successful at their jobs are those who are able to foster a climate of cooperation and positivity among their team (Leithwood et al., 2008).

(iii) *Developing contingency plans:* Principals should always be prepared for the unexpected by creating a plan of action. When a crisis or other unforeseen occurrence occurs, it is important to have a strategy in place on how to handle it (Gurr & Drsydale, 2020). Staff, students, and other stakeholders can work together to create such plans. Strategies for remote learning, other teaching methods, and processes to ensure the safety of students and staff during a crisis are all examples of what might be included in a set of contingency plans.

(iv) *Working with other institutions and groups:* School leaders can collaborate with other institutions and groups to share knowledge and resources, create cooperative plans, and create resilient educational systems. This collaboration serves as a mechanism for the enhancement and equity of educational opportunities (Muijs, 2010).

(v) *Ensuring the well-being of their staff:* This may entail offering staff members the resources and assistance they need to cope with stress and preserve their resilience in trying circumstances. Pollock (2020) indicated that the COVID-19 pandemic changed the dynamics of a principal's role in a school. One of the roles of a school principal is to consider the well-being of their students and teaching personnel.

In conclusion, principals have a significant impact on the success of educational institutions. School principals can help ensure that education continues to be a priority even in the face of adversity by making plans for what might happen, encouraging creativity, establishing solid connections, placing a premium on the well-being of their personnel, and working with others. However, leadership in a resilient education system is more about the leader's reactions in crises and their adaptation to changes (Odhiambo, 2014). In other words, resilient leaders overcome obstacles using their attitudes, values, and actions (Makoe, 2023). The role of the school principal as an educational leader in a resilient educational system is critical. It is indicated that many changes happen because of legislative and regulatory policies, changes in the population, changes in the economy and culture, changes in the curriculum and technology, and changes in response to immediate crises or challenges in a school system. At this point, the school administrator is responsible for fostering and supporting resilient behaviors within the teaching staff (D'Anca, 2017). According to Turk (2018), resilient principals are characterized by these five characteristics: self-awareness, awareness of others, mindfulness, optimism, and empathy. With the help of these elements, principals can implement coping methods and resilience-building mechanisms in their schools (Halevi & Schechter, 2023).

According to Reed (2018), resilient leaders have three characteristics: Resilient thinking, resilient capacity development, and resilient action skills. With the help of these skills, it is possible to create an action plan in emergencies and to continue education. For these reasons, finding resilient leaders among the representatives in the resilient education system is considered necessary to ensure the continuity of education. Learning outcomes are one of many focuses of the resilient education system. As highlighted by the UNESCO Deputy Director General for Education, Stefania Giannini:

> When we talk about building the resilience of the education system, it's not just about policy development and service delivery, it's also about schools providing emotional support to teachers and students and building students' own resilience. (UNESCO, 2020)

As Giannini stated, developing and implementing hybrid or distance education within a plan to create a flexible and resilient education system is not enough. It should also provide emotional support to parents, teachers, and students (UNESCO, 2020). UNESCO's recommendations for building a resilient education system for emergencies are summarized in Tab. 2.

Tab. 2. Building a Resilient Education System (UNESCO, 2020)

Prevention	**Preparation**	**Response**
o Data collection on school absence during emergencies o Creating a healthy & hygienic school environment o Developing policies for disadvantaged regions o Training teachers about preventive health services	o Performing analyses to identify student risk groups o Ensuring that schools make disaster management plans and that their basic needs such as nutrition and vaccination are provided o Establishing mental health support systems for students and teachers	o Ensuring continuity of education for all students o Providing professional psychological support to teachers o Making the academic calendar allow schools to open and close o Procurement of hygiene and safety products

As shown in Tab. 2, a resilient education system includes the prevention, preparation, and response steps. The prevention step includes absenteeism data and preventive health services. The preparation step includes risk group analyses and disaster plans. The response step includes the continuity of education and the flexibility of the academic calendar. At each step, providing training on cleaning, hygiene, and emergencies to the representatives during the training and offering professional psychological support is critical.

Here, the primary and secondary elements that make up resilient education, also defined as sustainable education, are summarized. The following section will present some resilient education practices from around the world.

Implementation of the Resilient Education System around the World

Resilient education has been a common ground for educators. For example, in the 1840s, continuity in education was ensured through education by snail-mailing. More recently, in 2016, schools were closed in Sierra Leone due to the Ebola virus, and education continued with a 30-minute radio broadcast (The Center for Global Development, 2020). With COVID-19, hybrid education and distance education have become indispensable for resilient education. With the closure of schools, decision-makers had to turn to distance and hybrid education modalities, requiring the use of distance education strategies. Today's distance education strategies can be listed as TV, radio, online courses with interactive participation, and mobile applications. The use of these strategies differs by the socioeconomic status of each country. Fig. 1 shows data on income level and the use of distance education strategies.

In distance education, each country takes some measures as best as it can. In Fig. 1, distance education strategies used to implement these measures are shown. As can be seen, high-income countries use more internet-based applications in distance education to provide resilient education. On the other hand, only 9% of upper-middle-income countries use internet-based applications, and 31% use at least three distance education strategies together. In low-income countries, at least three distance education strategies are used together. As such, high-income countries rely on their internet infrastructure, and students in these countries do not have problems accessing education. For example, whereas university students in Italy do not have a problem accessing distance education

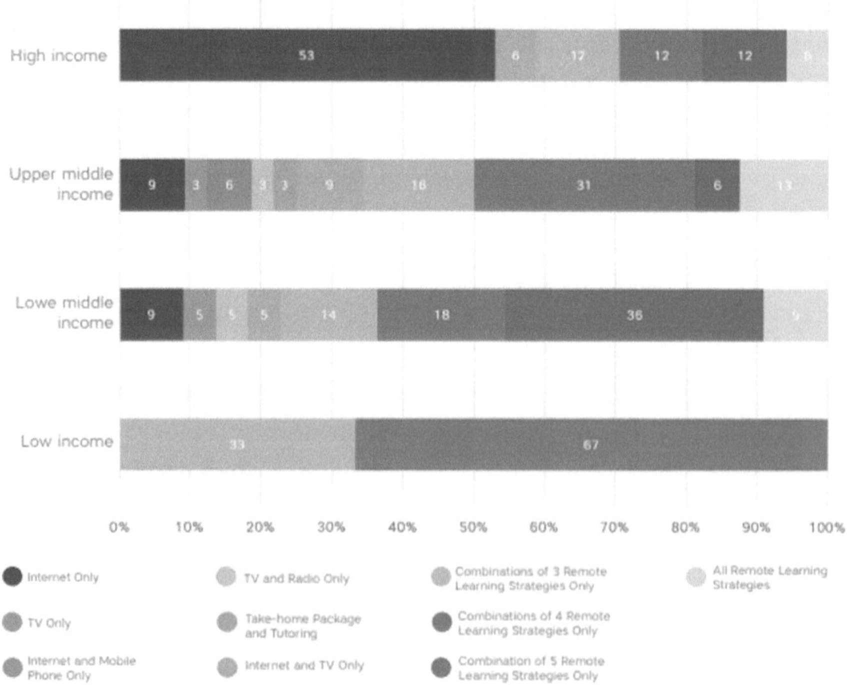

Fig. 1. Use of Distance Education Strategies by Income Level (Munoz-Najar et al., 2021)

(Bayrak & Önal, 2021), university students in countries with lower income levels have problems accessing distance education (Blaskó & Schnepf, 2022).

In addition to technological infrastructure and socioeconomic problems, the psychological consequences of the quarantines create a further source of concern for some countries trying to minimize them. For example, China is one of the countries where quarantines have been imposed for the longest time due to COVID-19. After months of school closures, China prepared a back-to-school support package and sent online surveys to teachers, students, and parents. In this way, China established regular communication with teachers, students, and parents early in the pandemic, eliminating concerns about returning to school and ensuring that it was safe to return to school (UNESCO, 2020).

This section summarizes how resilient education systems have been implemented in the world in the recent past. In the next section, how the resilient education system is implemented in Turkey, especially after the COVID-19 outbreak will be explained.

The Resilient Education System in Turkey

For a long time, Turkey has been trying to reach individuals who have difficulty accessing education using distance education systems, especially in tertiary education. However, given that COVID-19 is a respiratory disease, distance education is now implemented at almost every educational level. According to UNICEF data, schools in Turkey remained closed for the second longest time among the OECD countries between March 16, 2020 and May 31, 2021 (UNESCO, 2021). A summary of the school calendar implemented by Turkey during the COVID-19 pandemic is presented in Tab. 3.

Tab. 3. Timeline of Education Actions during the Pandemic in Turkey (TUSIAD, 2021)

Date	Action
March 12, 2020	Schools were closed for two weeks and education was suspended.
March 23, 2020	Emergency distance education began (UNESCO). EBA TV started broadcasting. It was announced that distance education would be implemented via the EBA platform until the end of the term (Papuşçuoğlu, 2020).
May 1, 2020	The EBA Academic Support program developed for the 11th and 12th grades was opened.

Date	Action
June 1, 2020	Nursery, day care centers and private preschool institutions were opened.
September 21, 2020	Schools were opened gradually for preschool and first graders. Face-to-face training was held once a week between September 21st and 25th, which was planned as an integration week.
October 12, 2020	Education with fewer desks has begun for Grades 2, 3, 4, 8, and Grade 12. Fully face-to-face education began in village schools, special education schools and combined classrooms.
November 23, 2020	Distance learning was extended until January 22, 2021.
February 1, 2021	It was announced that all public and private kindergartens, primary schools, 8th and 12th grades were to start face-to-face education as of March 1, 2021.

As shown in Tab. 3, it was not possible to keep schools completely open due to the uncertainty of the pandemic, and decision-makers had to make different decisions according to the direction of COVID-19. When the recommendations of UNESCO (2020) on building a resilient education system are examined, it can be said that Turkey was not prepared for this process. It can be observed that instant decisions are made only according to the increase and decrease of COVID-19 numbers. Due to the student's inability to access the internet to attend classes. A resilient education system should allow every student to have access to education. However, some students in Turkey have had problems accessing education, especially during COVID-19, mainly because of the digital divide between households (TUSIAD, 2021). According to PISA 2018 data, 31% of the children who go to schools in the middle and low socioeconomic regions of Turkey have a computer and internet connection to study at home, while 83% of the children who attend the schools in the regions with high socioeconomic level do so (OECD, 2021). According to the Household ICT Survey of TUIK (2022), while the rate of having a mobile phone in households in Turkey is 99.2%, the rate of having a desktop computer is 15.8%, and the rate of having a portable computer (laptop, notebook or tablet) is 36.1%. PC is the most used tool while maintaining distance education in Turkey. However, considering the ratio of desktop computers or laptops in these data, it is striking that most of them do not have a computer that can access education services at home, which may cause problems in implementing the resilient education system.

The research carried out in Tukey shows that students, teachers, and parents cannot effectively participate in distance or hybrid education. The research findings show that the parents of preschool children find the EBA platform insufficient in content and think that it should be enriched (Baysal et al., 2020). While some parents emphasize that it is positive that different times of the day are planned for access to EBA, parents with children at different grade levels describe it as negative that first and second grade classes are held at the same time. Preschool teachers, on the other hand, report that they had communication problems, difficulties in classroom management, and technical problems (Misirli & Deli, 2022).

Studies with parents and teachers of children in elementary school came up with the same results. Parents reported having connection problems (İnci Kuzu, 2020), teachers stated that they had little student participation in their classes and had problems connecting to the internet, and parents could not provide adequate support to their children (Kızıltaş & Özdemir, 2021), while the subjects in the curriculum were explained only superficially (Yurtbakan & Akyıldız, 2020).

The situation in secondary education is similar to primary and preschool education. Negative aspects such as connection problems and low participation rates were emphasized in studies conducted in Turkey by asking the opinions of students, parents, and teachers. They were also emphasized at the secondary education level. As for higher education, it can be said that universities with open and distance education faculties, such as Anadolu University, Istanbul University, and Atatürk University, quickly adapted to the emergency caused by the pandemic. In contrast, other universities had difficulties in this regard (TEDMEM, 2020). Except for faculties such as medicine and dentistry, distance and hybrid education were carried out until the 2022–2023 fall semester (YÖK, 2020). Various studies carried out with university students, and faculty members have revealed that applying distance education in applied courses does not have positive results (Bayrak & Önal, 2021; Kırtak-Ad, 2020). Another study revealed that 63% of university students do not have an internet connection at home, one-third do not have devices such as computers or tablets, and a quarter of them cannot continue their education due to these reasons (Karadağ & Yücel, 2020).

In this section, the implementation of the resilient education system in Turkey is summarized by considering the recent past. The final section below concludes the chapter with some suggestions.

Conclusion

The resilient education system has become extremely crucial, especially after COVID-19. However, very few studies have been conducted on the subject in Turkey. In the international research literature, hybrid and distance education systems stand out as the basis of the resilient education system. Throughout history, emergency education plans have been implemented in emergencies such as natural disasters, pandemics, endemics, and wars. The most recent example of this emerged in March 2020, and almost every country in the world has tried to take measures in its own way.

Considering the data of organizations such as the OECD, UNICEF, and the UN and the results of academic studies, it can be said that resilient education is applied in theory in Turkey. However, it needs to be developed and planned at some point because the purpose of the resilient education system is to reach out to every student under all conditions. However, some disadvantaged students in Turkey from preschool to higher education are excluded from the distance education system due to the differences in socioeconomic level, region-specific conditions, lack of internet access, and the limited or no availability of technological devices such as desktop computers, laptops, notebooks, and tablets in the household. To address this problem, ministries and NGOs can produce solutions for issues such as the free supply of technological devices and free internet service, and the elimination of infrastructure problems.

According to the latest OECD data, the poverty rate for the 0–17 age group in Turkey is 23% (OECD, 2021a), which acts as a limiting factor on access to hybrid or distance education within the resilient education framework. In an effort to reach out to all students from all levels in Turkey, EBA TV programs aim to address this limitation. However, it can be argued that poverty and the number of siblings of children in the household are also important factors for a resilient education system because there may be conflicts in the lesson hours of siblings in different age groups, and they may not be able to follow these EBA lessons. Therefore, during distance education, opening centers where the students can access the infrastructure and technological devices in the neighborhoods where the families live is strongly recommended.

The resilient education system is not a system where the teacher tells and, the student listens and takes notes, as in traditional education models. On the contrary, as in modern approaches, it includes an ongoing teacher-guided process. However, the number of students per teacher plays a decisive role in this guidance. According to the OECD figures, the number of students per teacher in Turkey is 15. The OECD average of this number is 13 (OECD, 2021b). In higher education, the number of students per academic staff in Turkey is above the OECD average. While the number of students per academic staff is 15 in OECD countries, this number is 23 in Turkey, and such a high ratio makes it difficult to provide high-quality feedback. It is highly recommended that the need for teachers and academic staff be met immediately to reduce the number of students per teacher and academic staff so that resilient education can be implemented more effectively.

Although hybrid and distance education models are important for a resilient education system, face-to-face education should be continued in areas where hybrid and distance education models cannot be adopted since the most important purpose of this system is to include every student. Necessary precautions like quarantine should be taken in emergencies such as natural disasters and pandemics, and plans should be made so that no child is left behind in education.

Assessment has an important place in the resilient education system. This should be carried out not simply to grade students, especially in hybrid education and distance education processes, but to compensate for existing learning losses and developmental deficiencies. Various strategies can be used to perform assessments and evaluations, such as e-portfolios, objective tests, research articles, and online discussions (Simonson et al., 2012). These can be applied throughout the process within the resilient education system, or they can also be carried out in schools when face-to-face education resumes after short-term quarantines.

A resilient education system should also include emotional support for teachers, students, and families since problems such as anxiety and fear may develop due to unexpected emergencies. In order to cope with these emergencies during the COVID-19 process, the Ministry of National Education of Turkey prepared guides to support the psychological resilience of students and parents. In addition, a helpline was established to provide this support. In addition, 1,375 experts were appointed to the Guidance and Research Centers (RAM) in all provinces of Turkey to support the resilience of students and parents (Özer, 2020), which was a positive development in a time like COVID-19 that requires prolonged shutdowns and quarantines. In addition, providing psychological support services to teachers increases the quality of teaching activities as it has a positive effect on teachers' well-being during emergencies. To ensure the continuity of education in emergencies such as disasters that can deeply affect human psychology, not only families and students but also teachers should be supported to increase their psychological resilience.

References

Bayrak, H. U., & Önal, N. T. (2021). SWOT analysis of distance education based on preschool teacher candidates. *Culture*, *6*(12), 189–234.

Baysal, E. A., Ocak, G., & Ocak, İ. (2020). Parents' views of preschool children on eba and other distance education activities during the *COVID-19* outbreak. *International Social Sciences Education Journal*, *6*(2), 185–214.

Bethke, L., Bird, L., Sigsgaard, M., & MacEwen, L. (2015). *Safety, resilience and social cohesion* (Glossary of Terms, Issue 36). International Institute for Educational Planning. https://unesdoc.unesco.org/ark:/48223/pf0000234806

Blaskó, Z., & Schnepf, S. V. (2022). *Educational inequalities in Europe and physical school closures during COVID-19*. European Commission, Joint Research Center. https://knowledge4policy.ec.europa.eu/sites/default/files/fairness_pb2020_wave04_covid_education_jrc_i1_19jun2020.pdf

Bozkurt, A., & Sharma, R. C. (2020). Emergency remote teaching in a time of global crisis due to coronavirus pandemic. *Asian Journal of Distance Education*, *15*(1), i–vi.

Center for Global Development, The. (2020). *Containing the epidemic: Should schools close for coronavirus?* https://www.cgdev.org/blog/containing-epidemic-should-schools-close-coronavirus

Dabrowski, A., Nietschke, Y., Taylor-Guy, P., & Chase, A-M. (2020). *Mitigating the impacts of COVID-19: Lessons from Australia in remote education*. Australian Council for Educational Research. https://research.acer.edu.au/cgi/viewcontent.cgi?article=1033&context=learning_processes

D'Anca, J. (2017). *Mindset and resilience: An analysis and intervention for school administrator* [Doctoral dissertation, St. John's University].

Dweck, C. S., & Yeager, D. S. (2019). Mindsets: A view from two eras. *Perspectives on Psychological Science*, *14*(3), 481–496.

Frankenberger, T. R., Constas, M. A., Neson, S., & Starr, L. (2014). *Current approaches to resilience programming among nongovernmental organizations* [Paper presentation]. Building Resilience for Food & Nutrition Security, AddisAbaba, Ethiopia.

Gurr, D., & Drysdale, L. (2020). Leadership for challenging times. *International Studies in Educational Administration*, *48*(1), 24–30.

Halevi, L., & Schechter, C. (2023). Resilient leaders: Resilient determinants when entering principalship. *NASSP Bulletin*, *107*(1), 60–86. https://doi.org/10.1177/01926365231159070

Hall, S., & Villareal, D. (2015). The hybrid advantage: Graduate student perspectives of hybrid education courses. *International Journal of Teaching and Learning in Higher Education*, *27*(1), 69–80.

İnci Kuzu, Ç. (2020). The view of the parents on primary school distance education program (EBA TV) implemented during the COVID-19 pandemic. *Milli Eğitim [National Education]*, *49*(1), 505–527.

Karadağ, E., & Yücel, C. (2020). Distance education at universities during the novel coronavirus pandemic: An analysis of undergraduate students' perceptions. *Yükseköğretim Dergisi [Journal of Higher Education]*, *10*(2), 181–192. https://doi.org/10.2399/yod.20.730688

Kırmızıgül, H. G. (2020). COVID-19 pandemic and the resulting education process. *Eurasian Journal of Social and Economic Research*, *7*(5), 283–289.

Kırtak-Ad, N. (2020). Prospective physics teachers' views on distance learning. *Education and Technology*, *2*(2), 78–90.

Kızıltaş, Y., & Özdemir, E. Ç. (2021). Classrooms teacher's opinions on distance education process. *Electronic Journal of Social Sciences*, *20*(80), 1896–1914.

Leithwood, K., Harris, A., & Hopkins, D. (2008). Seven strong claims about successful school leadership. *School Leadership & Management*, *28*(1), 27–42.

Makoe, M. (2018). Using future research methods in analysing policies relating to open distance education in Africa. *Open Praxis*, *10*(1), 5–15. https://doi.org/10.5944/openpraxis.10.1.581

Makoe, M. (2023). Resilient leadership in time of crisis in distance education institutions in sub–Saharan Africa. In O. Zawacki-Richter & I. Jung (Eds.), *Handbook of open, distance and digital education* (pp. 547–562). Springer.

Mısırlı, Ö., & Deli, H. (2022). Preschool teacher evaluations of the distance education process. In A. Courtney-Dattola (Eds.), *Handbook of research on adapting remote learning practices for early childhood and elementary school classrooms* (pp. 407–422). IGI Global.

Mısırlı, Ö., & Ergüleç, F. (2021). Emergency remote teaching during the COVID-19 pandemic: Parents experiences and perspectives. *Education and Information Technologies*, *26*(6), 6699–6718. https://doi.org/10.1007/s10639-021-10520-4

Muijs, D. (2010) A fourth phase of school improvement? Introduction to the special issue on networking and collaboration for school improvement. *School Effectiveness and School Improvement*, *21*(1), 1–3.

Muñoz-Najar, A., Gilberto, A., Hasan, A., Cobo, C., Azevedo, J. P., & Akmal, M. (2021). Remote learning during COVID-19: Lessons from today, principles for tomorrow. *World Bank*. https://documents1.worldbank.org/curated/en/160271637074230077/pdf/Remote-Learning-During-COVID-19-Lessons-from-Today-Principles-for-Tomorrow.pdf

Naidu, S. (2021). Building resilience in education systems post-COVID-19. *Distance Education*, *42*(1), 1–4.

Odhiambo, G. (2014). The challenges and future of public higher education leadership in Kenya. *Journal of Higher Education Policy and Management, 36*(2), 183–195. https://doi.org/10.1080/1360080X.2014.884676

Organisation for Economic Co-operation and Development. (2021a). *21st-century readers, developing literacy skills in a digital world.* https://doi.org/10.1787/a83d84cb-en

Organisation for Economic Co-operation and Development. (2021b). *Poverty rate.* https://data.oecd.org/inequality/poverty-rate.htm

Organisation for Economic Co-operation and Development. (2021c). *Education at a glance 2021: OECD indicators.* https://doi.org/10.1787/b35a14e5-en.

Özer, M. (2020). Educational policy actions by the Ministry of National Education in the times of COVID-19 pandemic in Turkey. *Kastamonu Education Journal, 28*(3), 1124–1129. https://doi.org/10.24106/kefdergi.722280

Papuşcuoğlu, T. (2020). A discussion on the right to education: An analysis of the distance learning process on Bourdieu. *The Human Rights and Equality Institution of Turkey Academic Journal, 4*(6), 129–148.

Patterson, J. L., & Kelleher, P. (2005). *Resilient school leaders: Strategies for turning adversity into achievement.* ASCD.

Pollock, K. (2020). School leaders' work during the COVID-19 pandemic: A two-pronged approach. *International Studies in Educational Administration, 48*(3), 38–44.

Readiness and Emergency Management for Schools (REMS) Technical Assistance (TA) Center. (2019). *Preparing for floods at state education agencies.* https://rems.ed.gov/docs/SEAFloodFactSheet_508C.pdf

Reed, D. E. (2018). Resilient educational leaders in turbulent times: Applying the leader resilience profile to assess resiliency in relationship to gender and age. *Perifiria, 10*(2), 119–134.

Sergiovanni, T. J. (2001). *The principalship: A reflective practice perspective.* Allyn and Bacon.

Simonson, M., Smaldino, S., Albright, M., & Zvacek, S. (2012). *Teaching and learning at a distance: Foundations of distance education.* Prentice Hall.

Tarricone, P., Mestan, K., & Teo, I. (2021). *Building resilient education systems: A rapid review of the education in emergencies literature.* Australian Council for Educational Research. https://research.acer.edu.au/cgi/viewcontent.cgi?article=1004&context=int_research

Tekel, E., Öztekin Bayır, Ö., & Dulay, S. (2022). Teaching practicum during the COVID-19 pandemic: A comparison of the practices in different countries. *International Journal of Progressive Education, 18*(2), 71–86. https://doi.org/10.29329/ijpe.2022.431.5

Turk, E. W. (2018). *Principals' perceived relationship between emotional intelligence, resilience, and resonant leadership throughout their career* [Unpublished doctoral dissertation]. Widener University.

Türk Eğitim Derneği MEM. (2020). *COVID-19 sürecinde eğitim: Uzaktan öğrenme, sorunlar ve çözüm önerileri* [*Education in the COVID-19 process: Distance learning, problems and solutions*]. Türk Eğitim Derneği Yayınları.

Türk Sanayicileri ve İş İnsanları Derneği. (2021). *COVID-19 etkisinde Türkiye'de eğitim* [*Education in Turkey in the impact of COVID-19*]. https://tusiad.org/tr/yayinlar/raporlar/item/10820-tusiad-erg-covid-19-etkisinde-turkiye-de-egitim

Türkiye İstatistik Kurumu (2022). *Hanehalkı Bilişim Teknolojileri Kullanım Araştırması* [*Household information technologies usage survey*]. https://data.tuik.gov.tr/Bulten/Index?p=Hanehalki-Bilisim-Teknolojileri-(BT)-Kullanim-Arastirmasi-2022-45587

United Nations Educational, Scientific and Cultural Organization. (2020). *Building back resilient: How can education systems prevent, prepare for and respond to health emergencies and pandemics?* https://unesdoc.unesco.org/ark:/48223/pf0000375278

United Nations Educational, Scientific and Cultural Organization. (2021, May). *Education: From school closure to recovery.* https://www.unesco.org/en/covid-19/education-response

Van Der Graaf, L., Dunajeva, J., Siarova, H., Bankauskaite, R., & Research for CULT Committee. (2021). *Education and youth in post-COVID-19 Europe: Crisis effects and policy recommendations.* European Parliament, Policy Department for Structural and Cohesion Policies. https://data.europa.eu/doi/10.2861/794039

Yamashita, T. (2018). *Grit and second language acquisition: Can passion and perseverance predict performance in Japanese language learning?* [Unpublished master's thesis, University of Massachusetts Amherst].

Yüksek Öğretim Kurulu. (2020). *Üniversitelerde uygulanacak uzaktan eğitime ilişkin açıklama* [*Explanation on distance education in universities*]. https://www.yok.gov.tr/Sayfalar/Haberler/2020/universitelerde-uygulanacak-uzaktan-egitime-iliskin-aciklama.aspx

Yurtbakan, E., & Akyıldız, S. (2020). Primary school teachers, primary school students and parents' views on distance education activities implemented during the COVID-19 isolation period. *Turkish Studies, 15*(6), 949–977. https://dx.doi.org/10.7827/TurkishStudies.43780

Part 3: Leading People: Communication and Collaboration in a VUCA World

Dr. Aybiçe Tosun

Managing Cultural Diversity for Equity and Inclusion

Abstract

Cultural and social aspects of societies have changed significantly due to globalization, migration, environmental crises, and economic problems in recent years. Living in a diverse society presents a variety of difficulties; while it may provide an opportunity to get to know one another, it may also lead to conflicts among different groups. One of the areas most affected by cultural diversity in society is educational environments. It is evident that issues such as prejudice, exclusion, marginalization, peer bullying, and lack of knowledge about others have arisen in educational settings with cultural diversity. This study aims to guide educational leaders in designing culturally diverse educational environments and dealing with the problems that may arise from them. There are various stages involved in addressing cultural diversity in the classroom setting: identifying the issues that can be described as problems, creating an awareness of similarities and differences among different cultures, guiding students to accept others with respect and tolerance, and finally creating basic principles for living together in a diverse society. Students have a major opportunity to interact with and learn from one another when cultural diversity is taken into account in developing learning settings. All students benefit when diversity is encouraged, intercultural education is better understood, and instruction is delivered with cultural sensitivity. It is essential that we teach our students how to tolerate people who are different from them and adapt to a changing world because of the surge in cultural diversity.

Keywords: Cultural diversity, diversity in education, diversity in VUCA world, leadership in diverse schools

VUCA and Diversity

> *"It is time to parents to teach young people early on that in diversity there is beauty and there is strength."*
>
> *—Angelou (2014)*

The age of migration has changed the world in general and societies in particular. First of all, societies that receive migration for various reasons have become more diverse than the previous generation. In these new diverse societies, cultural pluralism and the situations that arise with it have a significant impact on societies (Haas et al., 2022). This superdiverse and super mobile world has manifested itself most sharply in workplaces, schools, and daily activities.

One of the most frequently used adjectives to define societies, workplaces, educational institutions, and the virtual world is "diversity." According to Gollob and Krapf (2008, p. 40), "diversity implies moving beyond the idea of tolerance and genuine respect for and appreciation of difference. It is central to idea of pluralism and multiculturalism." Diversity refers to the existence of variation and differences within a certain group or community, which may include variations in racial or ethnic background, gender, sexual orientation, age, religion, culture, socioeconomic level, educational attainment, and other traits that distinguish people from one another. Appreciating and embracing differences, as well as simply identifying and accepting them, can help people better understand and value other points of view and experiences by fostering an inclusive environment that promotes diversity. This, in turn, can result in more creativity, innovation, and productivity in a range of contexts such as companies, educational institutions, and local communities.

The acronym VUCA has been used to describe such a complicated situation and highlights how the world has become unpredictable. The rapid changes in the social structure of societies, the super-connectedness to the rest of the world that comes with globalization, the more visible individual and cultural differences, and the increasingly heterogeneous nature of societies have become a challenge, especially for the field of education.

Diversity and VUCA are related because both can aid people and organizations in navigating the difficulties of a VUCA environment. By embracing diversity, individuals and organizations can have access to a wider range of ideas and methods for problem-solving. As a result, they may be better able to cope with the volatility, uncertainty, complexity, and ambiguity of their environment. Diversity can support the development of resilience and success in a VUCA environment for both people and organizations.

Leadership roles have been changing with the new social context. Berkel and Manickam (2020) state that managing differences is one of the most important key roles of a leader and conflicts, divergent viewpoints, and other differences must be managed. By doing so, leaders should promote variety because it can help them deal with changes in the external environment, no matter how slight. Kok and Jordaan (2019) argue that diversity management, social cohesion, mediation and negotiation skills are needed in a VUCA world for people who have leading roles. Similarly, Barentsen (2020) focuses on leader roles in a VUCA world; it challenges identities in individual and group levels and the role of leader must be challenged at the same time. Leaders should communicate by fundamentally altering how they perceive their own leadership role and identity, by being open to close listening and learning from their shifting environments, and by promoting frank and open discussion that allows uncertainties and doubts to be voiced and taken into account.

This book chapter focuses on how to design and manage learning and teaching environments in the VUCA world where cultural diversity has become a prevalent phenomenon. It will cover the definition, scope, and importance of diversity, the competencies associated with diversity, and how to comprehend diversity in the educational environment. The chapter will provide some recommendations for teachers as designers and managers of educational environments, educational leaders, and educational policymakers.

Diversity in the Classroom/School

Diversity in education includes an inclusive and respectful outlook and it also related to social justice. It is a multi-dimensional concept that deals with cross-cultural skills, social justice awareness and diversity knowledge (Washington, 1996). Diversity emerges in different dimensions in everyday life. The level of visibility of these types of diversity varies from society to society (Fig. 1)

The factors influencing the increase and spread of diversity can be listed as social violence, crisis situations caused by ecological violence and the resulting migration. Especially considering factors such as climate and violence, it is possible

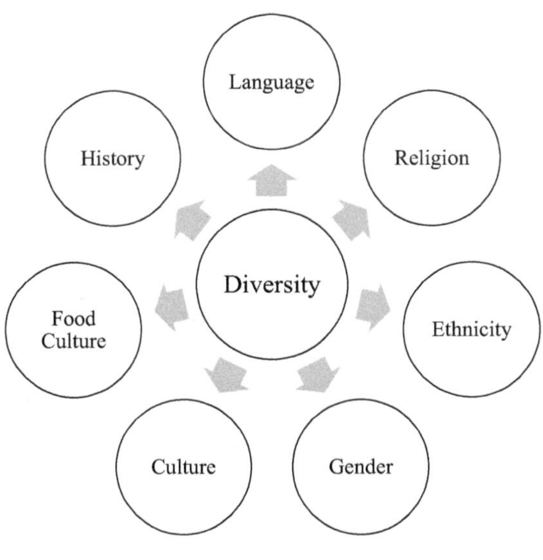

Fig. 1. Dimensions of Diversity

to say that the elements that constitute diversity are cultural productions. The crises arising from these can serve to show how fragile cultural productions are at the same time.

Especially the forced migration situation encountered in recent years has a great impact on the socio-cultural structures of societies. It is possible to say that one of the areas most affected by this situation is education. As a result of migration activities, the classroom environment becomes heterogeneous along with the rest of society. It can be considered that school is the place where individuals from different countries, cultures, religious and ethnic backgrounds most vividly observe the experience of living together in a new society. Learning environments can be seen as the place where important coexistence skills, such as self-expression, recognizing the other, discovering similarities and differences with others, and learning to cope with differences, can be developed. In this respect, learning environments should be considered as a fundamental space for the production and implementation of new methods and opportunities for the construction of a culture of coexistence. It will make a significant difference if educational leaders, such as teachers, trainers, and school principals, recognize this situation and get support from appropriate approaches and techniques for the integration of students, training of trainers, design of learning environments, and problem-solving.

Refugee movements due to internal conflicts, violence, environmental crisis, and other challenges have been one of the most important factors in increasing diversity in educational settings. According to UNHCR data, 51% of forced migrants are school-age children (UNHCR, 2017). Another UNHCR report shows that while only 68% of these children can attend primary school, 34% of refugees can attend high school and only 5% of refugees can attend university. In addition to these low rates, refugees are often an invisible element within national education systems (UNHCR, 2021). The low rates and opportunities for refugees to participate in education is one of the most important challenges related to refugee education.

Language barrier, peer bullying, discrimination, adaptation problems due to cultural differences, inadequate school capacities, shortcomings of educational leaders and teachers, financial constraints and lack of motivation in children are the main problems (ERG, 2017). One of the root causes of the problems refugee children face in accessing and receiving education is that educational environments are not designed to be sensitive to cultural diversity. Until recently, many European countries have not tended to develop their education systems to accommodate cultural diversity. For this reason, many teachers are not trained to design classroom environments with cultural diversity, especially as a result of migration. This situation results in the need to make some arrangements suitable for multicultural educational environments in countries that experience cultural diversity and to prepare teachers accordingly (Koehler & Schneider, 2019).

Forced migration constitutes an important dimension of cultural diversity. However, migration within or between countries for education or work also results in the increasing secularization of societies, with people of different religions living together and students studying together. Educational environments where differences between cultures are visible can be an opportunity for students from different cultures to get to know each other, but they can also create a ground for conflicts to emerge.

Cultural diversity resulting from phenomena such as rural-urban migration, global migration movements and refugees necessitates the redesign of educational and training environments. The general climate structure formed in the school environment with cultural diversity can be considered compatible with the VUCA world. As it is known, VUCA focuses on the fact that the world we live in is volatile, uncertain, complex and ambiguous. These characteristics that define VUCA are also reflected in learning environments where diversity prevails.

With cultural diversity, learning environments have experienced a transformation from homogeneity to heterogeneity and have become prone to volatility. This transformation brings with it the evolution of the educational atmosphere into a volatile structure with the coming together of different perspectives. This variability can be seen in the structure of the learning environment itself; the structure of the classroom environment changes in each new educational semester and a new structure may prevail compared to the previous semester in terms of experiencing differences. Changes in the student structure may also result in changes in the learning environment, learning program, learning approaches, methods and techniques used in the educational environment, materials and teacher training and competencies. One of the most important examples of variability in the field of education can be twenty-first-century skills; the learning skills that come to the fore in accordance with the needs of the current century have also changed. Learning outcomes that can be associated with diversity or the digital revolution such as collaboration, digital literacy and innovation are now on the agenda of education.

It is inevitable that uncertainty can be observed to a significant extent in environments with high variability. Along with cultural diversity, the fact that the educational environment or its elements are changing rapidly for various reasons brings with it uncertainty. Any change in the learning environment creates a complex environment that affects all members of the learning environment.

Complexity refers to the diversity within the classroom and the need to change accordingly. Finally, it can be observed that this whole process makes the educational environment ambiguity. Students' adaptation to the educational environment, their expectations from education, and their learning outcomes may become ambiguate, while teachers' management of the educational process may face problems in the same direction.

Understanding and Embracing Diversity in Education

Levine (1991) indicates that representation, support, integration, and multiculturalism are four main notions of diversity that can be seen in educational context. Every educational institution should clarify their agenda and develop long-term and comprehensive plans for future of this diverse world. These plans should include presidential leadership, frank and open discussions, and finally involvement of all community of education.

According to Sarin and Levanon (2022), education in a VUCA world is related to three important points as follows: (i) it contains conflicts between core values and practices, (ii) it ensures critical reflection and self-evaluation for whole process and (iii) it is usually related to interactions. Managing the VUCA world needs some key actions that leaders should take. Codrenau (2016) suggests that leadership in a VUCA world need to a comprehensive understanding about opportunities and challenges which come with the globalization and understanding, clarity and agility are the most important key actions for a VUCA world. Cultural diversity is one of the biggest challenges in a VUCA world for teaching and learning environments. Tuleja (2017) emphasizes that people need to be more culturally knowledgeable than ever to survive in the volatile, uncertain, complex, and ambiguous environment we live in today. Many of the communication difficulties based on by our various worldviews can be greatly reduced by learning more about ourselves as well as the various perspectives and values of others.

Diversity in the classroom is a significant challenge for students, teachers, administrators and other stakeholders. The ways in which this challenge is perceived and defined can be seen as a problem or an opportunity. From a positive point of view diversity in the classroom can assist to establish an inclusive and enriching learning environment for all parties in the system (Fig. 2). Research shows that students benefit from multicultural and diversity education and these contents enrich students' learning experience and enhance the understanding of differences (Piland et al., 2000).

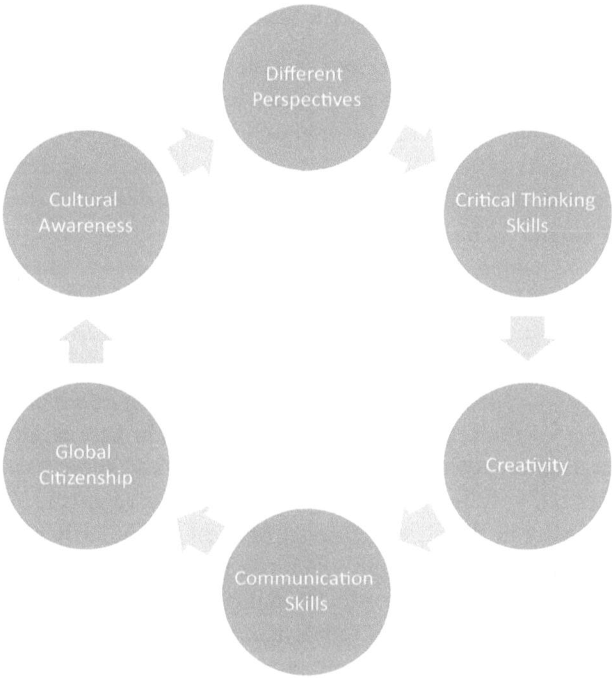

Fig. 2. Competences of Diversity in the Learning Environment

General competences that can be gained through diverse learning environments can be listed as shown in Fig. 2. In such classrooms, students from various backgrounds, ethnicities, and life experiences are grouped together, which can enhance everyone's perspectives and enable them to approach problems from multiple angles. Exposure to other points of view can aid in the development of critical thinking skills when students learn to assess and examine various arguments and ideas. Collaboration with people from different backgrounds can result in more original ideas as students are exposed to other ways of thinking and solving problems. Effective listening and speaking across cultural and linguistic gaps can help students grow as communicators by working with different peers. Students who learn to collaborate well with others from diverse backgrounds will be better equipped to succeed in their future occupations and as global citizens in today's connected society. Exposure to many cultures and viewpoints can aid in the reduction of prejudice and discrimination by fostering tolerance and understanding for various cultural practices.

How to Manage Diversity in the School/Classroom?

The most important stakeholders responsible for designing culturally diverse learning environments are teachers, educational leaders, and institutions and organizations responsible for educational policies. Although civil society organizations, community centers, and parents of students have significant influence in this process, this book chapter focuses on educational leaders, so details about these groups are not included.

Designing, managing, and resolving potential conflicts in volatile, uncertain, complex, and ambiguous learning environments where local and global variables are influential are challenging. To create a multi-voiced and fair learning environment where all stakeholders of the educational process are visible, a conflict-preventive approach should be adopted by being aware of the strengths and weaknesses of the process. For such an approach, first of all, similarities should be focused on without ignoring differences, and some basic principles should be adopted at different levels of education management. Managing the diversity that exists in the classroom environment is primarily possible by determining appropriate educational policies. Educational leaders are needed to support the integration of these educational policies into the school environment. Finally, there are principles that teachers, who are the administrators of educational practices in the classroom, should follow. The basic principles can be found in Fig. 3.

Policymakers

After teachers and educational leaders, the most important actors in educational processes are the people and institutions that make educational policies. Considering the practices related to cultural diversity and multicultural educational environments, it is not possible to achieve a useful result without these actors. Because it is a necessity for teachers, educational leaders and legislators to work together in the design of classroom practices and management processes in schools.

Training for Educators. One of the first questions that come to mind when thinking about multicultural educational environments is who will design these environments, who will implement these designs, and which approaches will be included in the process. Teacher training is essential in acquiring critical skills such as the general structure of multicultural education environments, their needs, and the resolution of conflicts that may arise in the process. Therefore, it is almost necessary that the first step regarding educational environments where cultural diversity exists is to provide training on multicultural education to all

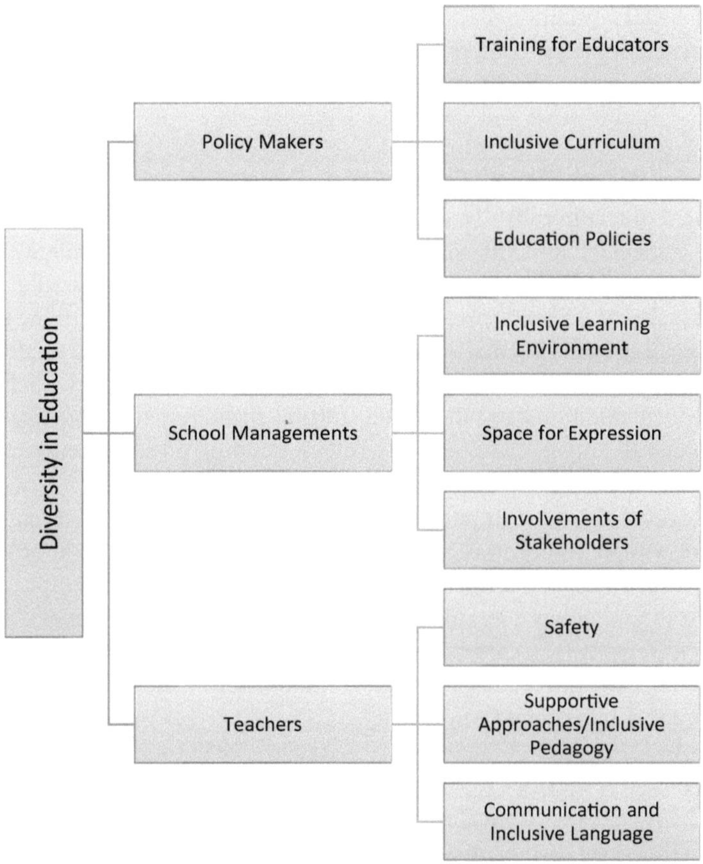

Fig. 3. Managing Diverse Classrooms

educators involved in the process, including teachers and educational leaders. These trainings should be included in both initial teacher training programs and ongoing professional development activities. After training the trainers, it is important to create opportunities for implementing new and diverse approaches. Meetings and workshops should be organized at national and international levels to share the results of the implementation with other educators and stakeholders.

Teacher training includes cultural competence and awareness. The first step toward a classroom environment with high cultural awareness is for the teacher to get to know their students. Getting to know students' cultural backgrounds, religious beliefs, traditional structures, sexual orientations, learning styles, etc., is the most important component of creating a learning environment that is sensitive to diversity. Knowing the students plays an important role in building a sense of mutual trust and can make students more open to the educational process.

Knowing the students brings respect for them, which is another important component of a culturally diverse learning environment. The teacher's recognition and respect for students' cultural diversity can be the key to students' respect for each other. At this point, it can be suggested that learning should explore students' cultural diversity, prepare assignments related to it, encourage them to design learning activities that can share their cultural aspects, and finally create extracurricular activities where cultural diversity can be presented. Such activities, in which students share their own cultural characteristics while gaining an awareness of the other, can play a key role in building positive communication between students.

Teachers are the most important elements of classroom practices and therefore have a high responsibility for cultural diversity. However, another very important factor related to the visibility of cultural diversity and respect for diversity in the school as a learning environment is educational leaders. It is impossible to separate the cultural sensitivity of educational leaders from teacher attitudes or classroom practices. For this reason, administrators' attitudes toward diversity are also extremely important.

Inclusive Curriculum. One of the most crucial factors for an educational environment that is sensitive to cultural diversity is an inclusive curriculum. There is a need for a flexible curriculum that allows different cultures to express themselves. At this point, two main directions emerge: first, direct outcomes related to cultural diversity and multicultural educational environments can be included in the programs. These outcomes can focus on skills, achievements, and values such as recognizing differences, understanding what is different, establishing a connection between oneself and the other, discovering similar and different aspects, coexistence, respect, and tolerance. Second, the curriculum should be flexible in terms of content and examples that are sensitive to cultural diversity while presenting different outcomes and allowing for the diversity that exists at the local level to be part of the curriculum.

Education Policies. One of the most important points that can be emphasized about education policies is the drawback of designing separate educational environments for students belonging to a different religion, ethnic group or culture. Such segregation of students is one of the biggest obstacles to the development of mutual recognition, understanding and coexistence skills. Cultural diversity can only become an asset if students are given opportunities to be together.

Participating in international studies is one of the most crucial ways to improve the effectiveness of actions carried out at the national level. Because of this, it is important to consider the experiences of other nations whose educational systems exhibit cultural variety while developing educational policies. Another important point is to work on ensuring that the diversity that exists and is visible among students also exists among educators. Multicultural educational environments should enable and encourage the cultural diversity of teachers and administrators.

School Management

School management is one of the most important actors in the education process. It both mediates between policymakers and practitioners in the implementation of educational policies and supports the design of multicultural educational environments by working with teachers. The main points concerning the areas of authority of the school administration are shared below.

Inclusive Learning Environment. School management should take a supportive approach toward teachers and other stakeholders in designing a learning and teaching environment that is sensitive to cultural diversity. It is crucial to involve teachers and students in making important decisions regarding cultural diversity, as this can prioritize the physical and psychological well-being of students and teachers and create a learning environment where cultural diversity is visible. Moreover, educational leaders should adopt a whole-school approach that involves students, parents, different stakeholders, and teachers.

Similar to teachers, it should be a priority for all educational leaders to become knowledgeable about intercultural pedagogy and familiarize themselves with different approaches. It is also essential for administrators to organize or direct teachers to attend training programs where they can develop their knowledge and skills in intercultural pedagogy. Teachers should be provided with necessary opportunities to apply different approaches that support cultural diversity in and out of the classroom. In parallel with cultural diversity in the school, it is crucial to have representations of cultural diversity among teachers and in school administration.

Space for Expression. Every student needs a space in the educational environment where they can express themselves freely. An important dimension of sensitivity to cultural diversity is to provide this freedom of expression to all students. At this point, it can be stated that the safe space in the classroom should be extended to the whole school. Teachers and educational leaders are expected to communicate and make decisions in harmony with each other to secure freedom of expression.

An important dimension of freedom of expression is the active involvement of students in decision-making processes. When new decisions need to be made for events and situations that have a significant contribution to the school culture, it is necessary to create a communication ground for students from different cultural backgrounds to reflect their own perspectives.

Involvement of Stakeholders. To support cultural diversity and multicultural educational environments, it is crucial for educational leaders to engage with local communities, civil society organizations and other stakeholders. This is effective in strengthening the dialogue between groups on the one hand and gaining a perspective on the needs of different groups on the other. Participation in national and international projects that will increase students' cultural awareness through stakeholders can be considered as an important opportunity. It is an important stage in understanding cultural diversity for students to experience multicultural educational environments that exist not only in their own schools but also in different schools and even countries. At this point, it would be appropriate to emphasize that even if there are no traces of cultural diversity in the student's educational environment, such projects are very important in expanding students' horizons.

Including communities and parents in the education process is a crucial step for integrated educational practices in a diverse society. Research shows that community engagement in teacher education programs for diverse educational environments is a significant part of the training (Schniedewind, 2001). The involvement of stakeholders in processes related to multicultural educational environments can also include educating the community, parents, and other families on a larger scale. These trainings should focus on knowledge, skills, and attainments such as recognition of differences, knowledge about multicultural environments, respect, tolerance, coexistence, and self- and other recognition.

Teachers

Some important points that fall under the authority of teachers as those who design multicultural education environments, work one-on-one with students, and manage the entire education process are shared below.

Safety. Environments where individuals feel physically and socially safe are also environments where they can freely share their thoughts. Such a learning environment implies a structure where all stakeholders are involved in the learning process and where different perspectives can be shared. To create such an environment, the roles and responsibilities of teachers, educational administrators, students, stakeholders, and parents involved in the educational process must be clearly defined. For this purpose, teachers and educational leaders facilitate healthy debates among diverse points of view and create opportunities to work together in classroom and extracurricular activities. For example, working in groups is always a beneficial way to learn from one another and it may create a solid space to practice freedom of speech and find a safe place to share. In addition, parents can contribute to the process by developing a harmonized approach and language with educational leaders. It is important to create an environment in which students can safely reflect their diversity and, if possible, to do so at both local and national levels.

Supportive Approaches. To ensure that all perspectives are given equal space in the educational environment and to address sensitive or controversial issues in a non-conflicting manner, it is important to adopt appropriate intercultural pedagogical approaches. While there are various approaches to multicultural education and teaching in diverse environments, this chapter will not delve into the details of each approach. However, some basic issues regarding the implementation of educational approaches need to be mentioned. First of all, teachers should be given the freedom to select and implement appropriate educational approaches for their classrooms. Each learning environment is unique, with its own strengths, weaknesses, and conflicts. Teachers are responsible for designing educational environments that meet the needs of multicultural education. At this point, Banks's (2004) suggestion can be considered. According to Banks, there are five basic dimensions of multicultural educational environments that educators can use as a guide when designing a multicultural school. These dimensions are content integration, knowledge construction process, prejudice reduction, equity pedagogy, and empowering school culture and social structure.

- *Content integration* is the degree to which teachers incorporate examples and material from various cultures and groups to illustrate important concepts, principles, generalizations, and theories in their subject area or field.
- *The knowledge construction process* has to do with how well teachers guide their students in comprehending, exploring, and identifying how implicit cultural presumptions, frames of reference, perspectives, and biases within a discipline affect how knowledge is created within it.
- *Prejudice reduction* refers to the lectures and exercises that educators use to assist children form favorable opinions of other racial, ethnic, and cultural

groups. This includes using multiethnic resources in a consistent and orderly manner as well as favorable depictions of the ethnic groups in the materials.
- *An equity pedagogy* refers that teachers should alter their methods of instruction in order to support the academic success of students from a variety of ethnic, cultural, gender, and social class groups.
- *Empowering school culture and social structure* is about including all parties in educational environment to all decision-making processes.

Another important point that comes to the fore in environments where diversity exists is making diversity visible in lesson plans. To increase cultural awareness, it is important to include examples or materials that reflect the differences that exist in the lesson plans. One of the most effective ways to increase cultural awareness is for students to learn through the social environment in which they live: their classmates, their cultures, or their history. Being flexible in designing classroom activities and giving students the space to express themselves freely can be considered an essential aspect of diversity-friendly approaches. To create an environment in which students can share their own perspectives and trust each other, it is necessary that both their communication among themselves and with the teacher is not restricted by rigid rules.

Communication and Inclusive Language. As communication is the most important element underlying all relationships, it is also the most important element of classroom environments where diversity is dominant. The teacher's preferred communication style in the classroom can set a meaningful example for students. For this reason, a language that affirms cultural diversity, is not marginalizing, avoids labeling individuals for their various characteristics, and does not characterize cultural differences as a separation is of great importance. At this point, it is important for the teacher, as the facilitator of the educational process, to guide students in maintaining their communication with each other in accordance with these general principles.

Conclusion

Education aims to give individuals a compass to find their way in the universe they live in. Finding one's way in the VUCA (volatile, uncertain, complex, and ambiguous) world, which affects our lives with cultural diversity, is first possible by discovering the fragility, strengths, and weaknesses of the world we live in. According to Morin (1999), an individual's inner balance and relationship with society are complex, and one of the main focuses of education should be to investigate the complexity of people and societies. The awareness of the common points of all humanity and the diversity of individuals and cultures is the most important step an individual will take to become a citizen of the earth. The emergence of the awareness emphasized by Morin will be possible with an educational administration that is suitable for a diverse society and does not push their individuality into the background.

The phenomenon of cultural diversity is ubiquitous across different sectors of society, including educational settings. It is evident that pluralistic societies require certain fundamental outcomes such as conflict prevention arising from cultural diversity, fostering a culture of reconciliation and peace, and promoting a cohesive coexistence of diverse groups. Achieving these goals would require a comprehensive understanding of the diversity prevalent in the educational environment and the subsequent development of tailored learning processes. Thus, it is essential to view cultural diversity not as a hindrance, but rather as an opportunity to cultivate mutual understanding and learning. Each encounter between individuals from different cultural backgrounds can be a valuable learning experience.

References

Angelou, M. (2014). *Rainbow in the cloud: The wit and wisdom of Maya Angelou*. Virago.

Banks, J. A. (2013). Multicultural education characterictics and goals. In J. A. Banks & C. A. McGee Banks (Eds.), *Multicultural education issues and perspectives* (pp. 3–24). Wiley.

Barentsen, J. (2020). Leading oneself in a VUCA world: Lessons from the field of religious leadership. In E. van Zyl, A. Campbell, & L. Lues (Eds.), *Chaos is a gift: Leading oneself in uncertain and complex environments* (pp. 211–224). KR Publishing.

Codrenau, A. (2016). A VUCA Action framework for a VUCA environment: Leadership challenges and solutions. *Journal of Defense Resources Management, 7*(2), 31–38.

Eğitimde Reform Girişimi. (2017). *Bir arada yaşamı ve geleceği kapsayıcı eğitimle inşa etmek* [*Constructing Living Together and Future with Education*]. https://www.egitimreformugirisimi.org/kapsayici-egitim/

Gollob, R., & Krapf, P. (2008). *Living in democracy.* Council of Europe Publishing.

Haas, H. D., Castles, S., & Miller, M. J. (2022). *Göçler çağı: Modern dünyada uluslararası nüfus hareketleri* (B. A. Evranos, Çev.). GAV Perspektif. (Orijinal çalışma basım tarihi 2013)

Koehler, C., & Schneider, J. (2019). Young refugees in education: The particular challenges of school systems in Europe. *Comparative Migration Studies*, *7*(28), 1–20. https://doi.org/10.1186/s40878-019-0129-3

Kok, J., & Jordaan, B. (2019). The metanarraphors we lead and mediate by: Insights from cognitive metaphor theory in the context of mediation in a VUCA world. In J. K. Kok, & S. C. van den Heuvel (Eds.), *Leading in a VUCA world integrating leadership, discernment and spirituality* (pp. 1–26). Springer.

Levine, A. (1991). The meaning of diversity. *Change: The Magazine of Higher Learning*, *23*(5), 4–5. https://doi.org/10.1080/00091383.1991.9939872

Morin, E. (1999). *Geleceğin eğitimi için yedi bilgi.* İstanbul Bilgi Üniversitesi Yayınları.

Piland, W. E., Hess, S., & Piland, A. (2000). Student experiences with multicultural education and diversity. *Community College Journal of Research and Practice*, *24*(7), 531–546. https://doi.org/10.1080/10668920050139695

Sarid, A., & Levanon, M. (2022). Embracing dualities: Principles of education for a VUCA world. *Educational Philosophy and Theory*, 1–12. https://doi.org/10.1080/00131857.2022.2162384

Schniedewind, N. (2001). Embracing the rainbow: An integrated approach to diversity education. *Multicultural Perspectives*, *3*(1), 23–27. https://doi.org/10.1207/S15327892MCP0301_5

Tuleja, E. A. (2017). Cultural intelligence in a VUCA world. In R. Elkington, M. van der Steege, J. Glick-Smith, & J. M. Breen (Eds.), *Visionary leadership in a turbulent world* (pp. 195–227). Emerald Publishing.

United Nations High Commissioner for Refugees. (2017). *Global trends: Forced displacement in 2016.* The UN Refugee Agency. https://www.unhcr.org/5943e8a34.pdf

United Nations High Commissioner for Refugees. (2021). *Staying the course: The challenges facing refugee education.* The UN Refugee Agency. https://www.unhcr.org/612f85d64/unhcr-education-report-2021-staying-course-challenges-facing-refugee-education

van Berkel, K., & Manickam, A. (2020). *Wicked world complex challenges and systems innovation.* Noordhoff.

Washington, S. J. (1996). Diversity education for professional practice. *Journal of Physical Education, Recreation & Dance, 67*(2), 42–44. https://doi.org/10.1080/07303084.1996.10607201

Dr. B. Çağla Garipağaoğlu

Leading by Becoming a Learning Organization in the Age of Quantum

Abstract

Senge (1990) states that the fundamental requirements of a learning organization require acting with moral purpose and mastering the following five disciplines: mental models, personal mastery, systems thinking, and shared vision. He also identifies 11 Laws, or truths of the fifth discipline/system thinking that can be used as a guideline helping educational leaders create a more hopeful future for children. Moreover, he introduces seven learning disabilities that largely go undetected in organizations and argues that an organization can become a learning organization only if it can take the necessary steps to overcome them. In order to navigate through the obstacles that so often clog and contaminate the educational system, the educational leaders and the teachers should (1) master the five disciplines of a learning organization, (2) internalize the merits and implications of 11 Laws, or truths of the fifth discipline, (3) be aware of learning disabilities and (4) tactfully deal with those learning disabilities that most often go undetected in the organizations by taking "system thinking" principle into consideration. By drawing upon the most influential work of Peter Senge (1990), this paper outlines us how to be a "system thinker" to successfully sail through rough waters of the quantum age for educational leaders.

Keywords: Quantum age, learning organizations, systems thinking, five disciplines

The Five Disciplines of Learning Organizations

Mental Models

Mental models are described by Senge (1990) as "our deep-seated, personally constructed stereotypes, generalizations and assumptions from which we make sense of the world." Most of the time, we do not even notice we possess them or know that they actually have an impact on our decisions. Mental models may also lock us into familiar ways of thinking and acting. Not only do we all create mental models, but we also share them. This is actually how an entire organization may have shared mental models which directs the organizational strategy and the way of doing things in the organization. Mental models also present some significant obstacles to our creativity and innovation by restraining us from learning. So, the only way to overcome the obstacles presented by mental models is to be aware of them and train our minds to act accordingly.

Senge (1990) states that people and organizations are all under pressure to decide quickly and find quick solutions to fix the problems immediately. This is actually why rather than thinking thoroughly about a problem and using our time to examine the facts and the conditions thoroughly we are so tempted to use their mental models to make judgements and dismiss problems. However, using these mental models as a shortcut in our decision-making processes without giving much thought to the facts and conditions not only may lead to incorrect conclusions, but it may also draw us into conflicting situations with others who may think otherwise. Senge (1990) argues that since the subconscious mind can deal with complex problems more swiftly than the conscious mind, all we should do is to train our subconscious mind, be aware of our mental models and how those models affect our decision-making. This understanding simply reminds us about the power of self-fulfilling prophecy theory (Merton, 1948). When we believe in our own power, this power eventually helps us realize our vision. The opposite is also true. If we believe in our own powerlessness, this time, this belief will hold us back from realizing our vision. The same principle goes also for our mental models. When we insist on believing in our entrenched mental models, our mental models will eventually lead the way and end up with realizing our prophecies.

Personal Mastery

According to Senge (1990, p. 141), personal mastery is "the discipline of personal growth and learning" and "the basic tenet of organizational learning that the other disciplines are built on top of it." Personal mastery is about becoming committed to figuring out one's personal vision and to be lifelong learner for a sustainable future. People who are skilled at personal mastery are generally good at generating a personal vision and evaluating their reality with respect to that vision (Senge, 1990). Since an organization's learning capacity can't exceed the level of its members, personal mastery happens to be one of the key points of organizational learning. The gap between reality and the future aspiration was called "creative tension" and considered as the leverage that propels people forward (Senge, 1990). Personal mastery is therefore about creating that "tension" and sustaining a peaceful life with and despite it (Senge, 1990).

People who are good at personal mastery often possess a special sense of purpose—a calling—that calls them for a service that is something larger than themselves (Meadows, 2008). They are also good at recognizing inaccurate assumptions and they are deeply inquisitive. They ask questions that no one dares to ask. Since they are always good at asking fundamental "why" questions, they can think out of box and find genuinely unique answers to the existing problems. They are resilient, learn from their own mistakes and regard the change as an opportunity (Meadows, 2008). They strongly believe in the importance of networks and people, yet they do not give up on their individuality (Senge, 1990). They have all the courage to stand up against the crowd, yet they are also still able to listen to other people's views respectfully and attentively (Covey, 1989). They are "systems thinkers" able to see themselves as a part of a larger system. Therefore, they have a unique sense of humility that makes them see themselves as a regular, ordinary but still significant co-player in a drama (Zohar, 2016). Therefore, they are able to live a responsible life by being responsive to each single moment and all that it inherits (Zohar, 2016).

Systems Thinking

Senge (1990) cites that learning organizations possess a holistic view called "systems thinking." Systems thinking stems from the system theory where each piece interplays with others interactively to form the "whole." Therefore, it grants a strong sense of belonging to the people in the organization. It also recognizes the importance of getting rid of the systems' hyperopia and developing an ability to see the bigger picture and the interplay between what might, at first, seem to be completely unrelated connectivity. It fosters a strong sense of belonging to something bigger than you. It also requires smooth and peaceful integration of diversity—valuing other people and unfamiliar situations for their differences, not despite them. System thinking relies on collective intelligence, meaning that a group of agencies are smarter than one or two intelligent agencies. This approach celebrates the subtle interconnectedness that grants living systems a unique strength—intelligence for free (Oswald & Mascarendas, 2018). It also requires a pair of eyes that are distant from the system and able to step away from the system itself to thoroughly examine and understand the interplay between the agents. Since a large, diverse group is better able to see a situation from all perspectives, "systems thinking" perspective advocates that a diverse group will be much better at identifying, framing, and solving a problem.

Team Learning

Team learning is not merely about improving team members' competencies, rather it is about the ability of "functioning as a whole." In a team, everybody must play a different role to satisfy the diverse needs of the organization and contribute to the accomplishment of the common goal. Just remember the Quidditch team in Harry Potter movies. Each Quidditch team has one keeper, one seeker, three chasers and two beaters. They all have different responsibilities, yet they have the exact same goal. And to achieve this ultimate goal, they all practice together, learn together, and set the strategy together. They learn to trust in each other and develop a team spirit so that they are able to "function as a whole." If a team fails to develop a strong team spirit, the problem of sub-optimization will surface at the expense of the ultimate goal. People will start to fear from turning their back on each other or venture on new journeys that will help them learn more and develop.

Senge (1990) argues that team learning is a compelling discipline that fosters and facilitates learning in an organization so that an organization can adapt and transform itself to accomplish its organizational goals in a fast-moving world. Senge (1990) notes that "considering colleagues as team members instead of competitors" is indeed the first step to initiate dialogues wherein people risk being vulnerable and express their free opinions without fear of being ridiculed. This is indeed the only reasonable way to create an organization where honest mistakes are forgiven, and it is safe to think and share freely (Senge, 1990). Otherwise, according to Senge (1990), no learning—individual or organizational—can be possible. When the working environment fosters "blaming" culture instead of a culture of "collaboration," learning something new or venturing into new territories will be impossible for the whole organization because people will expend all their energy just to preserve the status quo and protect their turf. So, team learning is only possible in a working environment where a culture of collaboration prevails instead of culture of rivalry and distrust.

Shared Vision

Senge (1990) emphasizes the importance of generating shared vision through togetherness and the interaction of all members of the whole organization. Although many leaders have personal visions, they often fail to make them a shared one in the false hope that others will adopt it with a simple snap of a finger, and this is indeed why visions fail: lack of a buy-in. According to Kouzes and Posner (2008), rather than envisioning exciting possibilities, the real challenge is to enlist others into a shared vision. Indeed, Kouzes and Posner (2008) notes that the most important attribute that stands out leaders from non-leaders is their ability to bring people together around a shared vision. And the only way to generate a shared vision and build a shared ownership of the vision in the enterprise is to let employees take part in the shaping process of the vision through "communication" and "co-creation." If people do not genuinely buy into the vision, it can only be reasonable to expect that their contribution to the professed vision will be very limited and merely superficial. On the other hand, if the employees share the same vision sincerely, they will perform the organizational tasks and fulfill their organizational roles just because they are willing to do so, but not because they are expected to do so, or they were told to do so.

11 Laws, or Truths of Learning Organizations

Today's Problems Come from Yesterday's Solutions

Every single "dilemma" or "dead-end discussion" we experience today owe its very existence to an intentional or unintentional step we took yesterday. In other words, decisions we individually or collectively make today often turn out to be tomorrow's troubles. It means that the mute discussions we are having today are likely the result of a series of decisions that are made for the problems that are no longer exists today and by the people that are probably no longer around. Therefore, the challenge is to avoid sowing the seeds of tomorrow's problems with today's solutions. To do that, all we must do is to train our minds to identify and comprehend the complex chain of causal factors that led to inherit the yesterday's problems and solutions that now consume our time and threaten our day-to-day businesses.

All we must do is to understand that shifting the burden where it does not belong may be the very source of a new and even a bigger problem. For instance, due to severe budget cuts, less funding might be allocated to education when we need to attract new and more investors for the country. However, this might be the very reason why new investors do not invest in our country where they cannot have access to a highly educated workforce and where the children of their employees will be forced into a substandard public education system. Thus, instead of focusing on short-term wins and interventions, it is better to focus on finding solutions by not shifting the burden where it does not belong. Astute problems solvers do not shift the problem where it does not belong, instead they simply invite all the stakeholder into problem-solving process and benefit from the collective intelligence to understand the root causes and find a solution that will not end up being the cause of bigger or worse problems.

The Harder You Push, the Harder the System Pushes Back

According to Senge (1990), well-meaning interventions in the current system will eventually call forth responses that might nullify any potential improvement in the system whether it is incremental or radical. It means that the greater effort you exert to produce a change or make an improvement, the more and more effort it will require. It is also called as "compensating feedback." Compensating feedback occurs when we do not notice that our well-intentioned interventions are contributing to our current problems while creating many others to deal with (Senge, 1990). As humans it is only natural to get exhausted and drawn into the process of compensating feedback and glorify the entailing suffering (Senge, 1990). When our well-meaning efforts to improve or change the current operations fail initially, we push harder and often expect that our heroic efforts will be enough to overcome all the obstacles in front of us. We all find comfort in sticking to what we know the best. Insisting on what we know as the best and clinging on the very system that creates the fundamental problem in the first place is indeed an indicator of non-systemic thinking—and what Senge (1990) calls the "what we need here is a bigger hammer" syndrome.

If you want to generate different results, you must face the fact that you probably can't get what you want with your current system. As famously pointed out by Einstein you cannot solve a problem with the same mindset that created it in the first place (Michalko, 2001). The leverage lies in designing new systems to generate new results, not in expecting already failed systems to create better results (De Bono, 1994). Push your system and the people in it as hard as you want and you might only get temporary results, but you will inevitably generate new problems and probably never understand why your hard work did not solve the problem but caused it to get even bigger. Unless you find out the root cause of your problem and address it properly, you are doomed to chasing your tail around the wrong learning loop. Organizations suffer from this syndrome when, for example, they begin to lose their popularity. When they begin to push new marketing strategies aggressively it often turns out that more money is spent on the marketing with merely a temporary pay back (Senge, 1990). We also suffer from a compensating feedback loop in education as well. For instance, when poor graduate performance getting started to be pointed out as the major problem in the industry, the typical reaction becomes to tighten the screws by raising the threshold for academic standards in the system of education. Unfortunately, by doing so, the compensating feedback gets into the play and this well-intentioned intervention merely cause teachers, students, parents, and principals become more and more frustrated and further increase demands for more stringent academic standards. So, when the system is no longer the right system, instead of forcing it to produce better results, we must change it altogether. Dakota tribal wisdom says that "when you're on a dead horse, the best strategy is to dismount. Of course, there are other strategies. But after you've tried all these strategies, you're still going to have to dismount." To sum up, for sustainable success, we will be better off when we abandon things that are no longer working.

Behavior Grows Better Before It Grows Worse

Symptomatic cures upon problems will make things look better for a while; but eventually, problems will arise again. Short-term and low-leverage interventions give temporary improvement at best, but never eliminate fundamental issues and problems. This is akin to treating symptoms rather than the disease. These underlying problems will tend to get worse, grow and spread in the long run. It won't be long enough before the problem comes back, or worse, it strikes again more forcefully and worse than the first time. Considering how frequently most positions are taken over by new people in organizations, someone else will be in charge of dealing with the new problem that should not exist at all in the first place.

In a typical example, a software developer may bypass the testing process to deliver a delayed project in time. From the management point of view, the developer might be praised because of the un-delayed deployment. However, over time, application failures because of lack of testing might give the developer a hard time at his performance appraisal—perhaps worse than the scenario where he would admit that the project was delayed. In the same line of thinking, when we might celebrate our short-term successes of saving money through the budget cuts on education, for instance. However, the mood of educators and students might naturally be affected by these cuts negatively. Consequently, the student and teacher morale may begin to falter leading to a decline in educational outcomes. When the frozen expenditure on education might relieve the financial pressures on the system temporarily, it may also trigger a considerable decline in educational standards. Just because it might be hard to recognize the cause-and-effect relationship between the two incidences at first sight, the underlying reason might be hard to detect. Yet, it must be noted that "the ignored problem becomes so massive over time … it eventually calls for a 'fireman' approach that can merely save the lives but cannot afford to save more."

The Easy Way Out Usually Leads Back in

This generally happens when we want to apply familiar solutions or "best practices" to our every single problem. Problems are part of our everyday life. So is problem-solving. What differs and what really matters in this regard is the fact that every individual is unique in their ability and approach to dealing with a problem. The biggest reason for never-ending problems in the corporate world is "firefighting" approach in problem-solving. When faced with a problem, our instinct tells us to clear out the problem as soon as possible. This is indeed why we tend to jump headlong into figuring out a solution that will fix the problem without analyzing facts thoroughly. Most problems stem from our intolerance to uncertainty. We should never forget the fact that the problems will always be there but how we handle them is up to us. When something works, we like to reuse it. The main problem with this solution-centric approach is that it is plainly flawed. Because it requires spending more time and energy on how right or wrong a solution is rather than understanding and analyzing the root cause of the problems thoroughly. Insisting on familiar solutions to solve our problems is pretty much like "using an old map to see a new land." It is simply irrational. Instead, we should consider all the possible options available regardless of how irrelevant they might seem now and make peace with the uncomfortable.

Kaplan (1964) called this "the law of the instrument." It is like giving a kid a hammer, and he will assume that everything needs pounding. As cited in Klatt (1999), Maslow (1966) explains this law with the following example: "If all you have is a hammer, everything looks like a nail. And eventually you start to treat every single problem as if it is a nail." An extreme example of this law is "a drunken man" who is searching for his keys (which he lost at some dark corner) under the streetlight just because it is more luminous or "a carpenter" who is trying different types of hammers to get a nail out of a wooden log when what he really needs is a pair of pincers. This is called a "comfort zone challenge" (Klatt, 1999).

However, according to Einstein, "smart people don't jump into the water without testing its depth," instead, they spend more time with a problem, and step away from it to understand it better (Michalko, 2001). The more we stay with a problem and spend time thinking about it, the better we become at solving problems. Neither staying with the problem longer nor thinking harder about the problem at length offers immediate pleasure or satisfaction to people. That's indeed why the problems become persistent and ubiquitous over time. We should all train ourselves to avoid the ease and comfort of the "easy way out" and keep an open mind about the problems by thinking imaginatively. Only then we can become better players in this fast-changing game of life.

The Cure Can Be Worse Than the Disease

Short-term fixes can be worse than the problem itself. Most employees spend their time creating and solving problems which should not exist at all. It is a waste of time and productivity. In the quest to be busy or hero of the day, people tend to create tomorrow's never-ending problems that should not exist at all. As we rely more on interventions, the more the system loses its ability to cure itself. The most dangerous and subtle consequence of solutions that do not synch well with the system thinking risks the need for more and more interventions. This is why poorly thought of solutions are not just ineffective. They are also worsening the situation as they foster more and more dependency for many more interventions. This is akin to the parable of "giving a man a fish" and "teaching him how to fish." If we are about to introduce an intervention into the system, then we must first guarantee that it doesn't lower the system's ability to cure itself by creating more reliance on new interventions. For instance, eliminating educational programs to lessen the pressure on the state budget not only disguises the need for more money, but it also weakens the entire education system by triggering a dismantling chain reaction. Therefore, the money we saved in the short term is likely to cost us more money in the long term. Likewise, alcoholism may simply start as a way to socialize or as a way to solve the problem of work-related stress. However, continuous drinking can lead to a much bigger problem in the long run: a diseased liver. In the same line of thinking, shifting the responsibility of teaching children from families to schools through public education triggered many more problems while solving many others. When the public education failed to shoulder the whole burden, the private sector emerged to share the burden, also causing many more problems in the long run. It must be noted that every solution designed to solve any problem might accompany a non-negligible potential to generate many more problems, eventually turning problems into massive dilemmas.

Faster Is Slower

The story behind this law is old: The slow turtle wins the race against the fast rabbit. Every system has an optimal and unique rate of growth that is far slower than the fastest possible rate. The fastest rate, in most cases, although it seems to be desirable, is not necessarily the optimal rate. When growth becomes too much to carry along, the system has a hard time coordinating itself, struggles to compensate for it by holding it up. All systems follow this rule inevitably. This natural resistance to growth can entail certain risks: for instance, because of raising costs, even the survival of the organization might be at risk. However, most of the organizations today will still prefer to push the growth as fast as possible, without considering the consequences of it because most of them will be blindsided by the short-term success and reward that accompany the principle of faster is slower. Therefore, the smartest thing to do would be removing the obstacles blocking the systems' veins instead of pushing for more and more growth.

For instance, the farther our educational system is detached from the realities of your young people, the more likely the leaders in the education system will be blindsided by distorted reports the sycophantic underlings about the wisdom and efficacy of current educational reforms designed to increase academic standards. They will neither care or notice that increasing academic standards is more than what our current industrial education system is able to absorb, nor it is what our educational system actually needs. Educational leaders are having a hard time understanding why their house of cards came crashing down. Even worse, it will probably be too late, or it won't even matter when they understand it. Just because their quick fixes would be rewarded with promotions, or their lifetime won't be long enough, they won't even be around to experience and see the full effects of their quick solutions. Effective and long-lasting solutions take time. So, instead of going after short cuts leading to poor solutions just because they are easy and quick, we should be wary of the quick fixes and search for the right solution by taking time.

Cause and Effect Are Not Closely Related in Time and Space

We all think that the apple does not fall far from the tree. This is why we have been conditioned to see problems in snapshots. Most of us search for the source of the fire nearby when the fire reaches out to us. And we conveniently accuse the nearest person to the fire when the fire started. Most of us lack the skills and potential to see the underlying reasons behind the problems we encounter daily. This is why we keep on encountering similar problems repeatedly and suffer from all their symptoms. Unless we address the real causes of our own problems, we keep on struggling with the numerous versions of the same problem. Some say that we are forced to take the same exam over and over by the universe, until we pass it. It is also a doctrine of reincarnation/rebirth which is central to Buddhist teachings. Since the causes and effects are not closely related in time and space, but rather distant, it's surprisingly difficult to understand how seemingly unrelated events stemmed from the same root causes and end up with a chain reaction. It is surprising that the underlying causes of most of the complex problems can be traced back in some unsuspected part of the system, in the very distant past. Despite all the difficulty of discovering the truth about the underlying causes of our problems, we must train ourselves to acquire the necessary skills and patience to "go upstream" and find the real causes of our complex problems.

Small Change Can Produce Big Results, but the Areas of Highest Leverage Are Often the Least Obvious

So far, it has been stated that obvious solutions don't always bring the best results. However, sometimes, a very small, and insignificant change can cause a considerable improvement. Seemingly insignificant but well-placed actions at the right place, at the right time in the system can produce the most solid solutions and can bring about a massive change in the most desired direction. This "systems thinking" concept is called "the law of leverage." The challenge is that the "right place" and "the right time" is not always obvious. Most often, it is even rather counter-intuitive. The areas of high-leverage are usually highly non-obvious as they are not "closely related in time and space" to the symptoms of the problems. There are no simple guidelines to find high-leverage areas. But as the first step, one can start learning to step back from the problem and see the big picture to figure out the underlying repeating patterns.

Some say that our schools are not up for the real purpose of the education system. But it is not true at all. Given the original goal of producing factory-ready workforce for the industry, our schools work perfectly well. The problem is that the goal is no longer the same. We no longer need factory workers, what we need are innovators, entrepreneurs, and creative talents. So, we have to redesign our education system to produce the new intended outcome. Yet, the education structure is massive, and it is really hard to discover where real the leverage points lie. So, the only way to overcome this difficulty is to take advantage of the collective intelligence of a large and diverse group of stakeholders for the benefit of the entire educational system.

You Can Have Your Cake and Eat It too, but Not All at Once

Systems thinking claims that some problems are not problems at all. Once we shift our mindset from a "snapshot" mode to a "process" mode, the problems might seem different. If we change how we approach the problem, we may not even see it as a problem any longer. For instance, many production managers would think that "low cost" and "high quality" are rigid "either-or" choices and cannot exist together. While it may be true at a "snapshot" mode of thinking, it might be wrong in the "process" mode of thinking. For instance, once IKEA placed lower prices for its products compared to a year earlier and stated that this happened because they were able to lower the prices thanks to the high volume of sales, less costly and more efficient new production methods. More often than not, it might be true that we must choose one or the other, but sometimes or rarely we can have both. And the real challenge or leverage lies in detecting those rare times when we can have both and act upon it and use the opportunity when it comes along.

The same line of thinking suggests that, although it was ineffective, costly, time-consuming, and inefficient to integrate technology into teaching and learning process at the early stages of technology, later it became much easier and less costly over time. And in the end, it proved to be very fruitful both economically and academic-wise. So, you can have your cake and eat it too although it may not be possible at once. All you need is to invite diverse stakeholders into the process of imagining viable solutions and allow time for those solutions to work.

Dividing an Elephant in Half Does Not Produce Two Small Elephants

All systems have "integrity." Therefore, it is not possible to take a part of a system and expect it to function well in isolation. Organizations, like living organisms, have integrity and their behavior affects and are affected by "the whole." In order to comprehend most managerial dilemmas or design a proper strategy to deal with them, we must first figure out how the system is functioning as a "whole" and why the problems arise or where the underlying problem lies. To illustrate this principle, Senge (1990) shares a Sufi tale about four blind men touching various parts of an elephant. Blind man touching the ears of the elephant declared that the thing he was touching was the "fan." The other one touching the "trunk" of the elephant declared that it was a "snake." The other one touching the "leg" of the elephant declared that it was a "tree." The last one touching the "tooth" of the elephant declared that it was a "spear." They were all right about how those parts were really felt, yet none of them were able to guess that it was an "elephant." Failure to comprehend the system as "a whole" might cause countless problems.

The education system might be considered as one BIG elephant. Therefore, the best approach to deal with the system of education would be to employ more eyes looking at the massive elephant from different perspectives instead of examining the fragmented parts independently and individually. However, Senge (1990) warns us about the difficulty of seeing the whole in practice by pointing out the fact that the organizations are designed in a way to hinder people from noticing important interactions.

There is no Blame

We are all connected. And the "systems thinking" teaches us that "there is no outside" and "there is no separate other." We are all part of a single system together with the very cause of our own problems. We all play our unique roles in how things turn out. Yet, we still tend to deny that fact and blame others when we fail. Although it is hard to admit that together with our enemies, we indeed co-create our problems. Yet, once we embrace this fact, it is going to be much easier to fine-tune and exploit the system for our own advantage.

Senge (1990) suggests that the solutions lie in constructing healthy relationships with the very people we blame for our own problems. The challenge is to seek comfort in inviting those who we see as enemies into the problem-solving process. Although many may fear that this will rock the boat, on the contrary, this can merely provide the much-needed ballast to "keep an even keel" in a storm. Considering the fact that the way all organizations were designed, all the jobs were defined, and we have all been taught to think and interact, it is not surprising that most organizations learn poorly (Senge, 1990). Despite the best efforts of all well-intended people, all those facts create some fundamental learning disabilities which would prevent an organization from becoming a learning one.

Learning Disabilities that Can Hinder Personal and Organizational Learning

I Am My Position

This is the tendency to limit our responsibilities and impact with the boundaries of our positions. This might cause people to consider themselves successful even if the general outcome is not successful. When asked what they do for a living, most teachers would answer this question in a way to specify their expertise, for example, "I am a math teacher, or I teach math," only a few say that "I'm an educator," and most of them would never relate to the purpose of their enterprise in which they take part. It is because most of them think that they have minimum or no impact at all in their enterprise. Most people merely deal with the problems that—they think—fall into the limited confines of their positions. However, as they tend to narrow down their impact and define themselves with their position only, it becomes harder for them to act together to achieve the shared vision. When people in organizations focus only on their position, they wrongfully assume little sense of responsibility for the overall outcomes, especially if the outcomes are disappointing. All too often, people will just assume that "someone—but not themselves—screwed up." And this eventually lessens the entire organization's learning capacity.

Hamel (2001) argues that in the long run, the most vital question for an enterprise is "not what you are or what your position is, but what kind of a role you may play in the future with your existing potential in your own industry." According to Hamel (1990), an enterprise can control its future only if it understands "what kind of a potential it carries" and able to define itself out of its own positional boundaries. A threat to the future is narrow-minded managers who define their mission with their current assets and operations and formulate their vision as the extension of it. It is important to note that creating the future is not the job of geeks or top management, it is the job of the whole enterprise. Creating a future requires "thinking out of box." Therefore, the tendency of teachers and educational leaders to narrow down their responsibilities and impact within the narrow boundaries of their job descriptions is the real threat to the education system, and it is what is really plaguing the entire system.

The Enemy Is Out There

People tend to find something/someone to blame when they fail. Teachers blames ministry for crowded school curriculum and low pays, parents blame school administrators for poor standards and lack of accountability, the ministry blames teachers for poor performance, students blame teachers for pushy deadlines, and everybody blames the ministry for its choice of central examination in student's replacement. This disability can also be considered as a by-product of "I am my position" disability. When people refuse to see how they contribute to their own problems, their tendency to find a scapegoat for their failures increases. This is indeed the outcome of not perceiving the world as a system but a series of unrelated and isolated parts. When people don't consider the whole system and the way it works, it becomes very easy to find something or someone else to blame when they fail or come across some unpleasant consequences. When we concentrate merely on our job descriptions, and do not understand how our own personal choices and decisions may play a role beyond the boundary of our job descriptions, when the consequences of our own actions hit back to haunt us, we just assume that these problems are created by others as if they are not our own creations. The only way to avoid this syndrome and overcome this disability is to assume responsibility for our actions and realize that the effects of our actions/decisions or lack of our actions/decisions may well-extend the boundary of our positions. Unless we hold ourselves accountable not only for the tasks described in our job descriptions but also for the long-run consequences of our own actions, no personal and organization learning can be fully experienced.

The Illusion of Taking Charge

It happens when "reactivity" is perceived mistakenly as "proactivity" (Senge, 1990). "Taking charge" is the term to take initiative and act early before things get out of control. However, this proactive approach can sometimes lead to bigger or worse problems. Consider a school outsourcing its hardware and software development needs to another company. A proactive school administrator can foresee that the development requirements will rise dramatically in the following years and can decide to train their own IT teacher to save money. This may sound like a good idea. However, while celebrating the illusion that the school administrator "took charge of the situation," s/he will miss the fact that this inexperienced IT teacher will probably be inadequate for the school's hardware and software development needs and just because his workload doubled, he will also end up with a poor performance in his/her teaching performance and burnout eventually. In this case, lack of systems thinking will cost the school greatly in terms of both hardware and software quality and educational effectiveness. It is important to note that all too often, pro-activeness is called as "reactiveness in disguise" (Senge, 1990).

Fixation on Events

You would think you would see a gorilla wherever it is or would not miss aliens crossing the street right in front of you at any time. De La Torre and García (2018) suggest that this may not be the case at all. They argue that we may not even notice them because of a "cosmic gorilla effect." This syndrome is best illustrated by an experiment where the researcher told the subjects to watch two teams throwing a ball between them and to count how many times the ball has been thrown. Since the people were so fixated on counting the number of times the ball has been passed between the two teams, almost half of them failed to see a cavorting man with a gorilla suit crossing the playground. This experiment proves that fixation on events can distract people to the point that they can even notice a massive gorilla right in front of their eyes.

We all have such fixations when we are performing our day-to-day businesses, and this eventually and surely stifles our creativity. This syndrome can be best depicted with a picture of Nero playing the fiddle alone and calmly while Rome burned. While he didn't fiddle during the fire, he is surely the symbol of leaders who misunderstood priorities in a crisis and focused on the exact wrong things at the exact wrong time. This disability is also self-evident in the failure story Kodak and Polaroid. While Silicon Valley embedded cameras in phones, Kodak and Polaroid preferred focusing on better film and paper. Although the executives understood that going digital was the "new normal," they feared to cannibalize their core film sales and failed to initiate the transformation triggered by the digital disruption.

In education, healthcare, agriculture, communications, genetics, politics, marketing, and commerce, the status quo is upended by new players who approach the game differently (Christensen, 1997). The incumbents try to solve the same old problems better with excellence and passion while the new players make the old problems irrelevant and concentrate on solving brand-new problems. Social and technological change often combine to radically rearrange our world, our ways of thinking, our ways of doing and our ways of life. Therefore, if you're doing everything by the book and merely focusing on excellence in existing operations with passion, then when faced with a disruptive innovation coming from a tiny competitor, or a startup company in the garage, you will most probably live the very same destiny of Kodak and Polaroid.

The Parable of the Boiled Frogs

This disability is about the famous story of the boiling frog who cannot notice gradual changes in the temperature of the water until it is too late. Much like a frog in a pot of water gradually heated to boiling would relax into drowsiness and die without putting up any fight to survive, most of us cannot even think of jumping out of the pot before it is too late and would wait to be boiled to our deaths without resistance. However, if you throw a frog into already boiling water, it will jump out of the pot at the instant. The reason for this phenomenon lies within the nerve system of the frog. A frog's anatomy is built to make sure that it will react upon sudden changes. If its environment changes slowly, the frog is not able to notice it. This parable happens in organizational life as well. Most of us fail to see our own demise if it is coming slowly enough. It is indeed a natural result of short time orientation and lack of systems thinking. According to Christensen (1997), when a company finds a successful business model, management exploit it to its fullest, and each passing year, it becomes more difficult for the company to imagine a time that its business model has reached to the end of its life cycle. If you don't want your current business model to determine your company's life cycle, you must remember that most successful organizations fall as they fail to notice the need to regenerate their core strategies, and instead act as maintenance engineers, and not architects (Hamel, 2001).

Many executives think of "new ideas" as unnecessary distractions that interfere with the day-to-day businesses of an enterprise (Catmull, 2014). Unless these "new ideas" generate short-term wins, they would fail to receive any attention. This is why Catmull (2014), Co-founder and President of Pixar, calls them "ugly babies"—not mature, not attractive, and requires protection until they are mature enough (2014). Christensen (2014) thinks that the end begins for organizations when those babies are left alone without protection and argues that "coziness is not a hotbed for innovations" and without innovations and new ideas, all organizations are doomed to die or be killed by others. To some extent, success is an innovation killer and for successful organizations, the failure of innovation is primarily due to fear and fixation on the past success. If failure is not an option, the ability to learn is also not part of the organizational culture. Success makes people stick to a recipe. All too often, the end begins when companies start to perpetuate the past practices reflexively—by over-valuing conformance more than creativity and by transforming long-tenured nostrums into sacred truths (Hamel, 2001). When the management overinvest in what they already have, as opposed to what they may have potentially, they start to bring about their own demise gradually and eventually. According to Christensen (1997), "It is the necessity that fosters innovation, not the coziness."

The Delusion of Learning from Experience

It is certain that one of the best learning methods is learning from direct experience. If we do it by ourselves, we learn better. The most important prerequisite of this method is getting feedback. We need to get feedback to "learn" whether we did it right or not. However, when we are acting in an open system, it is very hard to see the global results of our decisions and actions. The most important consequences of our actions usually lie in the distant future or a distant part of the system, this is why we may not necessarily and always experience the full effects of our actions. The things we learn in such an environment can be right at the time being, but they can also be wrong in the future. Just because something seems to work right in our own learning horizon, it doesn't mean that it is always right. For example, real consequences of decisions in R&D will be visible after sales. Just because a hair dryer seems to weigh less after changing the type of plastic of the cover, the R&D scientist shouldn't hurry to "learn" that the new plastic type is better. After a year, the market share of the hair dryer can decrease because "it feels cheap." Just because some consequences of our actions lie beyond the current limits of our awareness, we may not see the consequences of our actions fully. No other explanation of this syndrome is as vivid and clear as this "Those who live by the sword … get shot by those who don't …" (Hamel, as cited in Solomon, 2016). The original biblical version, "those who live by the sword, die by the sword" on the other hand, suggests that the way you lived your life would be the way you end it. However, Hamel's modification suggested something else. Yesterday's solutions and path dependency might pave the way to death! Path dependency refers to the tendency to rely on past practices, decisions, and actions rather than current conditions. The path dependency theory can save time by providing a shortcut, but it does not necessarily give you the best or optimum solution. In other words, sticking to a traditional path is not always the ideal strategy because it blinds you to the other available options. Therefore, if the ultimate goal is to survive and prosper, it is not recommended to depend on historical preferences.

The Myth of the Management Team

The responsibility of the management team is to make sure that the organization's pathway is clearly defined, departments are interacting correctly, and the common goals are achieved. However, all too often, because the managers tend to care more about protecting their turf and avoiding anything that will make them look incompetent, unfortunately over time "this team" will function in a way to minimize any chance for collective or individual learning (Senge, 1990). Meanwhile, the teams will also pretend that every single member of the organization rally around the shared vision and supports organizational strategy design by team collectively. In order to highlight the seeming cohesiveness of the team, they will also seek ways to suppress oppositions and disagreements of people with serious reservations. This attitude will gradually cause people to avoid stating their oppositions publicly and decisions will most likely be watered-down by the trade-offs.

At a more senior level, employees learn that their ability to influence others and garner support is more important than their mastery of an operational skill. As they progress to climb up the ladder, they realize that "better performance" does not necessarily imply for greater rewards on the part of the employee but "compromising" in order to look good to the top management—regardless of the benefit of the enterprise—may prove to be more profitable at the individual-level. That is a critical crossroad both for the enterprise and the individual as the employee may have to make concession on his or her professional integrity for the sake of workplace politics (Dvir, 2015). With the expectation for promotions, higher pay or the possibility of vast layoffs looming ahead, many people may join the political dance. In the end, a dedicated employee who once was committed to deliver value to the enterprise may turn into an ordinary bureaucratic beast paying its lip service to the top management. It does not mean that politics should be avoided by pretending that it does not exist. Politics is inevitable with people. The problem starts when politics becomes the main course and people start solely to focus on allegiance, politics, and personal agendas (Dvir, 2015). This toxicity manifests itself in the forms of capitulations, protocols, hung decisions, compromises on personal and professional integrity and a fear culture (Dvir, 2015). This is why, at the coziness of the top, no one really cares about the enterprise, or its goals. They are there to ensure that no one threatens the status quo. All they care about is to manage up, look good at the eyes of the bureaucratic guards and to make sure that there's enough protocol to cripple any change effort. This culture damages organizational learning as it paves the for mediocracy instead of meritocracy and pushes the true talents and the dedicated workforce out of the system.

Conclusion

By drawing upon the most influential work of Senge (1990), this paper teaches educational leaders how to be a "system thinker" to successfully sail through rough waters of the quantum age. It should be noted that in the age of quantum, no problem is strong enough to put us into a trouble, or put an end to a system, it can only force us to change or change the rules of the game. What does really matter is the choices we make and the attitude we choose to adopt. Change is absolutely inevitable but neither uncertainty nor the challenges we may face can pose any danger by itself. As long as we train our minds to embrace change and adopt the right attitude at the face of uncertainty, we may then adapt our systems to benefit from what the future might bring. All we need to do is to admit that everything is interconnected and there is no separate other. Instead of trying to isolate ourselves from our problems with an instinct of self-protection, we should remind ourselves that such an effort is not only useless but dangerous. When we have all the intelligence and capacity to leverage all these interconnectedness, we just push ourselves to get out of our comfort zones.

Senge (1990), thus, advocates that in order to navigate through the rough waters that so often clog the educational system, the educational leaders and the teachers should (1) master the five disciplines of a learning organization, (2) internalize the merits and implications of 11 Laws or truths of the fifth discipline, (3) be aware of learning disabilities and (4) tactfully deal with those learning disabilities by taking "system thinking" principle into consideration.

References

Catmull, E. (2014). *Creativity, Inc: Overcoming the unseen forces that stand in the way of true inspiration.* Random House.

Christensen, C. M. (1997). *The innovator's dilemma: When new technologies cause great firms to fail.* Harvard Business School Press.

Covey, S. R. (1989). *The seven habits of highly effective people*. Simon and Schuster.

De Bono, E. (1994). *Parallel thinking: From Socratic thinking to de Bono thinking*. Penguin Books.

De La Torre, G. G., & García, M. A. (2018). The cosmic gorilla effect or the problem of undetected non terrestrial intelligent signals. *Acta Astronautica*, *146*(5), 83–91. https://doi.org/10.1016/j.actaastro.2018.02.036

Dvir, Y. (2015). *Don't rock that corporate boat*. Culture is King. https://cultureisking.org/2015/10/08/dont-rock-that-corporate-boat/

Gleick, J. (2011). *Chaos: Making a new science*. Open Road Media.

Hamel, G. (2001). Leading the revolution: An interview with Gary Hamel. *Strategy & Leadership*, *29*(1), 4–10. https://doi.org/10.1108/10878570110367141

Kaplan, A. (1964). *The conduct of inquiry: Methodology for behavioral science*. Chandler Publishing.

Kiel, L. D., & Elliott, E. W. (1996). *Chaos theory in the social sciences: Foundations and applications*. University of Michigan Press.

Klatt, B. (1999). *The ultimate training workshop handbook*. McGraw-Hill Professional.

Kouzes, J. M., & Posner, B. Z. (2008). To lead, create a chared vision. *Harvard Business Review*, 20–21. https://hbr.org/2009/01/to-lead-create-a-shared-vision

Meadows, D. (2008). *Thinking in systems: A primer*. Chelsea Green Publishing.

Merton, R. K. (1948). The self-fulfilling prophecy. *The Antioch Review*, *8*(2), 193–210.

Michalko, M. (2001). *Cracking creativity: The secrets of creative genius*. Ten Speed Press.

Oswald, A. J., & Mascarenhas, S. J. (2018). A systems thinking approach to understand the challenge of corporate ethics in the turbulent markets of today. In A. J. Oswald & S. J. Mascarenhas (Eds.), *Corporate ethics for turbulent markets: The market context of executive decisions* (pp. 79–117). Emerald Publishing. https://doi.org/10.1108/978-1-78756-187-820181004

Senge, P. M. (1990). *The fifth discipline: The art and practice of the learning organization*. Doubleday.

Solomon, Y. (2016). *Un-Kill creativity: How corporate America can out-innovate startups*. CreateSpace Independent Publishing Platform.

Zohar, D. (2016). *The quantum leader*. Prometheus Books.

Dr. Ceyda Aysuna Türkyılmaz

Marketing and Risk Management for Educational Leaders in VUCA

Abstract
Risk management is one of the most important processes affecting enterprises that leads them to success and sustainability. The importance of the concept has increased with the emergence of VUCA that makes environments volatile, uncertain, complex and, ambiguous. In other words, nowadays how to manage risks in VUCA world became more precise than ever. Considering the role of education sector in a country's economic and social development this chapter focused on risk management for educational leaders in VUCA. Chapter begins with a definition of risk and goes on with risk management process and strategies. In the later parts, enterprise risk management and enterprise risk management in VUCA world were defined. After discussing about the relationship between marketing and risk management in VUCA, suggestions were offered for the educational leaders.

Keywords: Risk, risk management, enterprise risk management, marketing

The Concept of Risk

In the relevant literature there are several different views about the etymology of risk (Althaus, 2005). One common view states that the word is originated from the Latin word "risicum" meaning "the one that cuts or the cut off." The term is associated with danger defining the cliffs that threatened the ships and sailors. Following this origin, the word took place in two different languages: Spanish as "riesco" and in Italian as "risco" (Liuzzo et al., 2014). Both words are associated with "danger." On the other hand, some other views assume that the word "risk" is originated from the Arabic word "risq" (Outreville, 1998) meaning "anything that has been given to you by God and from which you gain profit" (Althaus, 2005).

Aven (2012) points out that there is not a common and agreed definition of the concept of risk. He reviewed the literature and classified the risk definitions under nine different categories as: Expected value (loss), probability of an undesirable event, objective uncertainty, uncertainty, potential/possibility of a loss, probability and scenarios/consequences/severity of consequence, event or consequence, consequences/damage/severity of these and uncertainty on objectives. Two important issues draw attention from these categories of definitions. They are uncertainty and possibility of occurrence of an undesired outcome. This outcome can be related to different aspects in our lives such as finance, environment etc. Therefore, as stated by Fell et al. (2005) risk can be defined as "the measure of probability and severity of an adverse effect to life, health, property, or the environment." By citing Markowitz (1991), Renn (1992) points out that all risk circumstances focus on one common point: the distinction between the reality and the possibility. Accepting this distinction Renn (1992) defines risk as, "possibility that an undesirable state of reality (adverse effects) may occur as a result of natural events or human activities."

Considering the definition and meaning of risk, it is possible to say that risks can emerge in several disciplines in different ways. Althaus (2005) discussed that risk can be observed in disciplines such as logic and mathematics, science and medicine, social sciences (e.g., anthropology, sociology, economics, law, psychology, linguistics), history and humanities (e.g., history, the arts), religion and philosophy. The only difference between these disciplines is the form of the risk that occurs. Therefore, maybe the best way to classify the risks would be to use the sources that drive them. For instance, considering the sources from where the risks emerge, risks may be summarized as purchasing risks, financial risks, production risks, marketing risks, logistics and distribution risks, managerial risks etc. for an organization.

Risk Management: Process and Strategies

As can be seen from the definition and types of the risks, it is obvious that a rational and effective management of risks is essential. Otherwise, the possibility of the emergence of negative outcomes increases. Zio (2007) points out that in order to manage the risks a proper treatment of the uncertainties is required. Therefore, a need for a systematic management is necessary which leads us to the concept of "risk management."

British Standards Institute (2009) defined risk management as the "systematic application of management policies, procedures and practices to the activities of communicating, consulting, establishing the context, and identifying, analyzing, evaluating, treating, monitoring and reviewing risk." Aven and Vinnem (2007) state that, it is not possible to eliminate the risks totally, therefore it is essential to learn how to manage them. In a broader sense, they summarized the main purpose of risk management as to "ensure that sufficient measures are taken to protect people, the environment and assets from harmful consequences of the activities being undertaken, as well as balancing different concerns, in particular HES (Health, Environment and Safety) and costs."

Risk management process consists of several steps. In different sources the number of the steps included in the process may differ. The reason for this is in some sources the steps are combined and given a single name covering two or more steps. According to Outreville (1998), the risk management process can be classified in four steps: (i) Risk Assessment—consisting of risk identification and risk measurement, (ii) Risk Control—consisting of loss prevention and loss reduction, (iii) Risk Financing—consisting of internal financing, external financing or compensation and (iv) Implementation and Control.

In risk assessment step all the risks are identified. This step is very important because if the risks are poorly identified, the possibility of the emergence of negative outcomes would increase. So here in this step, potential risks are identified before they emerge. In the same step, under the measurement of risks sub-title, risks are evaluated which means that the potential size of the loss and its probability of occurrence is assessed. After this step, a decision should be given either transferring the source of the risk to a different entity, avoiding the risk, totally abandoning it or accepting it. Regardless of the firm to accept the risk or not, it must examine the possibilities to reduce the frequency of occurrence and to reduce the severity of losses that will occur. In other words, a kind of loss control is performed. In the next step that is risk financing, the necessary financial sources are found to overcome the possible losses as results of risks. And in the final step, the risk plan is implemented, monitored and controlled. These steps are summarized in Fig. 1.

Although risk management process and the steps included in this process are very important and vital, one should also take into account the strategies that can be used during risk management. When the relevant literature is reviewed, it can be revealed that there are different classifications of strategies used for risk management. According to Klinke and Renn (2001), there are three main and commonly used risk management strategies which are risk-based, precautionary and discursive strategies. These strategies use two main risk characteristics to assess and handle the risks. Risk-based strategies are appropriate for the risks where the assessment of certainty is adequate. Because of this reason in their following research Klinke and Renn (2004) might change the name of the category from risk-based strategies to science-based strategies. In precautionary strategies, the risks have a high degree of uncertainty. Therefore, precautions play a vital role pointing out a strict policy of containment, constant monitoring and continuous research and the development of substitutes (Klinke & Renn, 2001). Most important issue about these strategies is to be able to carefully monitor the signals and precursors of serious events. The last category of risk management strategies—the discursive strategies—are necessary especially if the potential

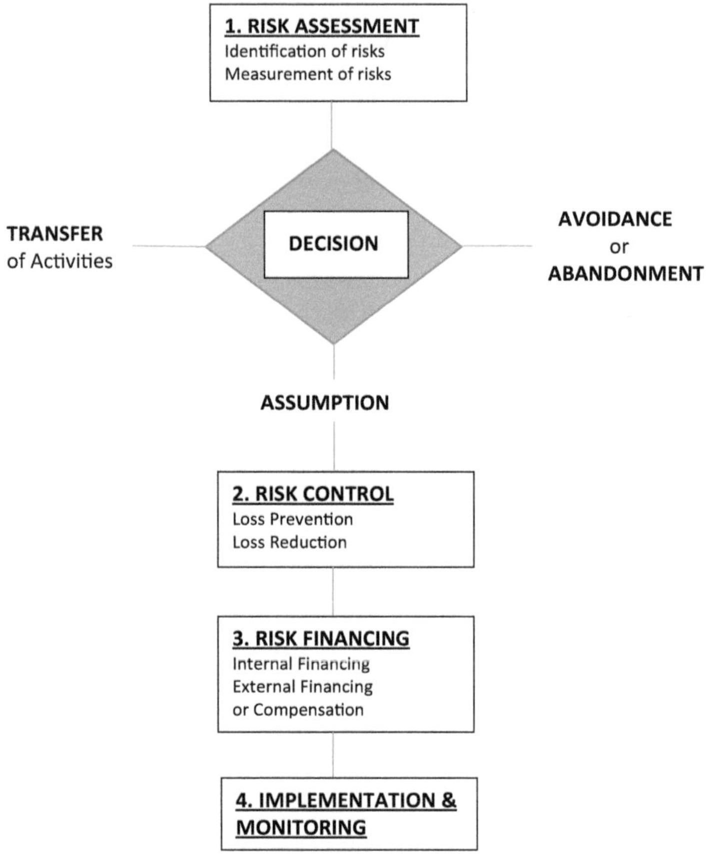

Fig. 1. The Steps in a Risk Management Process
Source: Outreville (1998).

for wide-ranging damage is ignored due to a delay effect. For discursive strategies, it is essential to create awareness, build confidence and offer trustworthiness (Klinke & Renn, 2001). Each of the risk management strategies and related actions are summarized in Tab. 1.

Tab. 1. Risk Management Strategies

Risk Management Strategy	**Damage Potential**	**Probability**	**Strategies for Action**
Science-Based	High	Low-Uncertain (depending on the risk category)	• Reducing disaster potential • Ascertaining probability • Increasing resilience • Preventing surprises • Emergency management
Precautionary	Uncertain	Uncertain	• Implementing precautionary principle • Developing substitutes • Improving knowledge • Reduction and containment • Emergency management
Discursive	Low—High (depending on the risk category)	Low—High (depending on the risk category)	• Consciousness building • Confidence building • Introducing substitutes • Improving knowledge • Contingency management

Source: Adapted from Klinke and Renn (2004).

In his later work, Renn (2008) identified risk management strategies by using the three main characteristics: complexity, uncertainty, and ambiguity. Depending on this point of view, risk management strategies can be summarized as follows strategies for coping with linear and routine risks, complex risks, uncertain risks and ambiguous risks (Renn, 2008).

Although these classifications are so precise and useful, maybe the best way to cope with risks in risk management is to customize the strategies for the unique requirements of the risks. Considering each risk's complexity, ambiguity, uncertainty, possible negative outcomes, coverage area will lead to a more suitable and rational climate to deal and manage risks.

Enterprise Risk Management

The definitions and the scope of risk and risk management reveal that risk and risk management are very broad concepts. In other words, it can be noted that risks may emerge in every aspect of our lives or in several disciplines. One of the areas that is interested in risk management is organizations. Power (2004) states that since the mid-1990s there has been a dramatical increase in the risk management practices of the organizations. Following this, there has been a growing interest, drive and enthusiasm in several industries and society to risk management in organizations (Aven & Vinnem, 2007).

Enterprise risks occur when there is a likelihood that the outcomes from a corporate strategy of a company differ from the ones that are identified in the corporate objectives and goals. In other words, it is the extent to which those outcomes fail to meet the objectives that were defined at the beginning of the planning period. Considering the definition of enterprise risks, enterprise risk management can be defined simply as; "the systematic and integrated approach to the management of the total risks that a company faces" (Dickinson, 2001). In a broader sense, the Committee of Sponsoring Organizations of the Treadway Commission defined enterprise risk management "as a process, effected by an entity's board of directors, management and other personnel, applied in strategy setting and across the enterprise, designed to identify potential events that may affect the entity, and manage risk to be within its risk appetite, to provide reasonable assurance regarding the achievement of entity objective" (COSO, 2004).

Two important issues can be drawn from these definitions. The first one is enterprise risk management which is the ongoing use of risk management practices in the organizations regardless of the industry that they are operating in. The second one is, enterprise risk management utilizes an integrated systems approach which means that the enterprise as a whole is taken into account when managing risks.

Dickinson (2001) points out that there are two main reasons of the need for enterprise risk management: (i) the increase in the high-profile company failures and preventable large losses, (ii) the growing importance of shareholder value models in strategic planning which pays more attention on risks to increase the value. Combining these two important reasons, it is possible to say that enterprise risk management is an essential process to obtain high levels of company performance, outcomes and returns.

Enterprise Risk Management in VUCA World

In the last section, the growing importance of risk management and the need for it in organizations were mentioned. This need becomes more essential in a VUCA World because of several reasons. Before explaining these reasons, it would be useful to remember the meaning of VUCA. VUCA is a word representing four important elements: Volatility, uncertainty, complexity, and ambiguity. These words can be explained as follows (Bennett & Lemoine, 2014):

- *Volatility:* Although information is accessible, and the circumstances are understandable there is an unstable and unpredictable change.
- *Uncertainty:* There is a lack of knowledge as to whether an event will have meaningful ramifications. Although causes and effects are understood, it is still unknown whether an event will create a significant change.
- *Complexity:* Change may not be involved but the situations are influenced by many interrelated factors making them difficult to understand.
- *Ambiguity:* Due to the lack of information causal effects are not understandable and clear and there is no precedent for making predictions as to what to expect.

VUCA world creates a more complex, rapidly changing, unpredictable environment which create many challenges, uncertainties and risks for the organizations. In order to reach the expected outcomes and returns organizations should deal with these challenges, uncertainties and risks in a rational and scientific manner. This can only be achieved by a properly performed risk management process. Tab. 2. focuses on the effects of VUCA world considering the macro-environmental factors such as technological, economical, ecological, socio-cultural, political and legal. The table summarizes the factors that trigger and have impact on risks in terms of volatility, uncertainty, ambiguity, and complexity. It also provides the consequences for risk management.

VUCA WORLD				
MACRO ENVIRONMENT		Trigger	Impact	Consequence for Risk Management
		Complexity	Uncertainty, Volatility, Ambiquity	Simplicity, Certainty, Stability, Uniquness
	Technological	- Digitalization& technological advances - Shortened product life cycles and increasing product diversity	-Acceleration of business operations&shortening of reaction time - Increasing uncertainty regarding product acceptance	-Determination of countermeasures and responsibilities immediately after the risk identification - Continuous risk monitoring enables instant counteracting measurements.
	Economical	- Globalization of markets - Expansions and M&A activities	- Interrelated international business relationships increase range of crisis. - Ascending number of transnational and cross-cultural corporate arrangements	- Holistic risk identification includes all relevant risks from business partners. - Centralized Risk Management Dept. is supported by decentralized risk managers and adequate IT-tools.
	Ecological	- Limited availability of resources - Global increase of natural disasters	- Use of alternative resources requires the reorganization of structures and processes - Serious unpredicted economical troubles endanger company existence	- Early identification of ciritical success factors and adequate alternatives - Preparation of emergency plans for existing threatening nature disasters.
	Socio-Cultural	- Demographic trends - Consumer-related, social trends	- Unclear customer behavior resulting from heterogeneous behavior etc. - Impacts from fast moving trends on product range policy	- Profound risk identification by means of adequate tools - Establishing appropriate risk response strategies.
	Political/ Legal	- Global political instability - Variety of highly differing laws and orders	- Impairment of business operations by labile political systems - Compliance violations due to variety of high differing laws and orders	- Extended risk monitoring of markets with unstable political systems. - Consideration of all regulations in the rish catalogue.

Tab. 2. The Effects of The VUCA World on Risk Management

Source: Adapted from Kuznik (2016).

Let's give some examples to explain Tab. 2. The globalization that spreads worldwide causes interrelated international markets and businesses which increase the likelihood of the occurrence of risks. To overcome this problem a holistic risk identification is required that covers all relevant risks from international business partners.

As another example from ecological environment can be related with the scarcity of the resources. In the recent years there has been a great emphasis on this issue due to the consciousness that people have. In this regard, resource-sensitive production and distribution processes should be reorganized due to new environmental regulations. At this point, risk management is essential to early identify the critical success factors and adequate alternatives (Kuznik, 2006). Going on with ecological environment, in the recent years there has been an increase in the occurrence of natural disasters. Unfortunately, one of them has happened in Turkey on the 6th of February this year. A very strong earthquake hit the southeastern parts of Turkey and nearly 11 provinces were affected very negatively resulting in thousands of deaths and injuries. In such situations risk management should cover extensive contingency plans to overcome the possible negative outcomes of these natural events that threaten the survival.

Considering the socio-cultural environment, it is possible to say that the changes in demographic and social trends result in unpredictable consumer behavior which means uncertainty for organizations and especially for marketers. In these circumstances, risk management is required in terms of risk identification and establishment of risk response strategies. The risk management process in VUCA world follows the same steps as in an ordinary setting mentioned in the previous sections. Kuznik (2016) summarized these steps as follows: risk identification, risk quantification (risk analysis), risk response (avoidance, risk reduction, limiting the risk/diversification, or acceptance of risks), and risk monitoring.

Marketing and Risk Management in VUCA World

Marketing is "satisfying human and social needs in a profitable manner" (Keller & Kotler, 2016). As the definition implies, the focal point of marketing is "individuals or specifically customers." Marketing's main purpose is to identify the customers' needs and wants and satisfy them in a profitable way so that both of the parties would gain benefit and result in a win-win situation. In order to achieve this objective marketing first identifies its target market. To identify the target market three main steps are essential: market segmentation, target marketing and positioning. After identifying the target market, marketing mix must be designed. Marketing mix or in other words 4Ps of marketing—product, price, place and, promotion—strategies play an important role in satisfying the needs and wants of the target market. Therefore, they should be carefully planned and customer-oriented.

These fundamental steps and activities in marketing mentioned above take place in marketing environment. The marketing environment can be analyzed under two categories: microenvironment and macroenvironment. The microenvironment consists of the actors closely related to the company and it directly affects the performance of the marketing activities. The company, suppliers, customers, competitors, marketing intermediaries, and publics are included in the microenvironment. On the other hand, macroenvironment consists of larger societal factors that affect microenvironment. Macroenvironment has an indirect effect on the performance of marketing activities through the microenvironment. Demographic, economic, natural, technological, political/legal, and socio-cultural forces are included in the macroenvironment (Kotler & Armstrong, 2016). It is obvious that both of the environmental forces have significant effects on marketing activities which influence the company's overall performance. The relationship between marketing and risk management can be explained by using two different points of view:

(i) Marketing-related activities or environmental forces can be the sources of risks.
(ii) Marketing activities, especially marketing research can be used as a risk management tool.

(i) Marketing related activities or environmental forces as sources of risks: In order to identify the risks and handle an effective risk management process it is essential and useful to detect the sources of the risks. The forces in micro and macro environments of marketing can be important potential sources for risks. These risks may affect the performance of marketing as well as the performance of the company. For instance, considering the macro environment.
Demographical changes such as changes in the age, income, ethnicity, family sizes of the population may create an uncertain marketing environment for the company due to the effects of these factors on consumer behaviors. These changes may create uncertain consumer behaviors. For instance, the migrations toward a country may affect the ethnicity of the population and create an unclear market segment for the marketers. In the recent years, lots of people from Syria migrated to Turkey. These people's preferences, needs and wants were undefined at the beginning and created a risk potential until they are detected, and marketing strategies are adapted for them.

The changes in the age distribution in a society may also create uncertainty. Marketers are still trying to identify the characteristics of Generation-Z which constitutes an important portion of the population in Turkey. Trying to understand this generation, their expectations etc. would help marketers to develop successful marketing strategies.

Technological improvements and developments caused the emergence of new ways to reach customers. Using internet and mobile applications online shopping has become more popular than ever. Although online shopping has several advantages for both companies and consumers, it resulted in a more complex and undefined consumer behavior that occurs on online platforms. This also creates uncertainty in the marketing decisions and activities.

Cultural issues are more important if the company is operating in international markets. Considering the elements of the culture it is possible to say that it has a huge impact on consumers' lifestyles and therefore, on their wants, preferences and consumption habits. Since each foreign country has its own unique culture, this may create an ambiguous market for the international marketers because some elements of culture such as religion may be difficult to interpret and take action toward.

Changes in the economic climate of the country may affect how consumers spend their incomes and plan their savings. Economic fluctuations may create a volatile market in terms of how consumers spend their money on shopping and how they prioritize their preferences. The political instability, war, crime, labor costs, corruption, nepotism, politically motivated violence, social unrest, transparency of fairness etc. may increase the perceived political risk in a country and therefore may create a risky market to enter.

Considering the micro environmental factors; the rapidly changing customer preferences, changes in the company's management, vision, mission etc., unpredictable actions of the competitors, changes in the attitudes of the publics' toward the company, problems that can be occurred in a relationship between suppliers (e.g., a delay in the lead times due to termination of a contract with a supplier) may create uncertain, ambiguous and volatile environment for the marketers.

The changes and challenges mentioned above may be the sources of risks. According to Konstantinovich et al. (2019), some risks are associated with wrong or incorrect actions of the company and called as subjective risks (e.g., economic and financial potential of the company). Those risks may also arise from marketing mix decisions such as a decision about the quality or features of the product, in appropriate price levels, or wrong communication messages etc. On the other hand, there are objective risks that emerge independent of the company's activities such as competition, inflation, and the economic and political situation in the country. In both risk categories, risk management is necessary for the organization with a different risk approach.

The pace of the changes in the micro and macro marketing environments mentioned above is greater than ever nowadays which increases the risk level and makes risk management more essential. Due to the rapidly and constantly changing market conditions, volatile, uncertain, less loyal consumer behaviors emerge which increase the likelihood of risk and cause marketing process to be more complex. These issues match with the elements of the VUCA world: volatility, uncertainty, complexity and ambiguity. In other words, in order to deal with the risks emerging from the marketing-related issues in VUCA world organizations and marketers should learn how to implement risk management process.

(ii) Marketing activities and especially marketing research can be used as a risk management tool: According to Konstantinovich et al. (2019) marketing is a very important risk management tool since marketing activities make a significant contribution to reducing risk, the likelihood of its occurrence. In this regard, marketing research may be very helpful and useful for marketers.
With the use of several data collection techniques of market research (e.g., surveys, in-depth interviews, observation, focus group interviews) it is possible to learn and gain deeper knowledge about the unknown or unpredictable consumer behavior, determine the optimum price level, forecast the customer demand correctly, gain information about the market conditions and trends etc. The information gained at the end of the market research would be useful to identify the risk or reduce it. Also, with a well-organized and performed market research it would be possible to convert risks into opportunities.

Other than market research, some other marketing activities such as finding a small market segment or market niche, using reputable distributors, applying advanced methods of sale and trade services can be used in risk management (Konstantinovich et al., 2019). Marketing research's role as well as marketing's role becomes more essential in terms of risk management in VUCA world where the markets are more volatile, uncertain, complex and ambiguous.

Suggestions For Educational Leaders in the VUCA World

As in most of the sectors, it is also possible to see the effects of VUCA World in the education sector. In other words, educational leaders are trying to continue their jobs and management styles in a more volatile, uncertain, complex and ambiguous environment than ever. Therefore, to be successful and sustainable it is essential to have knowledge on risk management techniques, procedures and strategies. Besides, understanding the importance of marketing and marketing management may be helpful for educational leaders surviving in VUCA World.

Educational institutions may encounter with several types of risks throughout their survivals or at least have a high possibility to face with risks. According to The Office of Chief Risk Officer at Stanford University (2023) some of the risks that an educational institution may face with can be strategic risks, healthcare risks, student and parents related risks, operational risks, financial risks and marketing risks. To reduce, detect and manage these risks as every organization, educational institutions and the individuals who take place in the management of these organizations require an effective and efficient risk management. Considering the role of educational institutions in a country's economic and social development, it is possible to state that for long-lasting and successful educational institutions risk management process should be carefully handled and adapted to the specific requirements of the institution.

Although risk management is very important and essential in educational institutions, this is not enough for success. Educational leaders should also understand the role of marketing in risk management process. For instance, in order to detect risk sources marketing research can be a very useful tool for them. By this way, risk sources may be identified in the external and internal environment of the institution and before risks occur, risk management plans can be organized.

In situations where risks occur marketing communication tools may be effective to give information to the target market (e.g., the publics, parents etc.). Especially if negative outcomes of risks emerge, they can be managed by an effective marketing communication strategy using tools such as public relations etc. Giving sincere and updated messages to the target groups can build trust and strengthen the institutions' corporate image and reputation. Transparency is one of the most important factors for the target market of an institution. In this regard, offering up-to-date and correct information with an effective communication can build long-term relationships with the institution and its target market (parents). In other words, this is one of the most important tools to create loyalty and therefore enough emphasis should be given on this issue.

Conclusion

This chapter aimed to draw attention to risk management in VUCA world for educational leaders. For sustainable success educational leaders should have knowledge on risk management and related concepts. In this regard, the role of marketing cannot be ignored. Marketing environment can be a very powerful source for risks therefore should be continuously traced. Marketing research is a very useful tool here to collect information about marketing environment. With the information gained from marketing information systems and marketing research educational leaders can plan and implement successful strategies and achieve sustainability. By drawing attention on risk management in VUCA, this chapter tried to offer suggestions for educational leaders with the use of marketing.

References

Althaus, C. E. (2005). A disciplinary perspective on the epistemological status of risk. *Risk Analysis*, *25*(3), 485–781. https://doi.org/10.1111/j.1539-6924.2005.00625.x

Aven, T. (2012). The risk concept—Historical and recent development trends. *Reliability Engineering and System Safety*, *99*, 33–44. https://doi.org/10.1016/j.ress.2011.11.006

Aven, T., & Vinnem, J. E. (2007). *Risk management: With applications from the offshore petroleum industry.* Springer Science & Business Media.

British Standards Institute. (2009). *Risk management: Principles and guidelines.* BS ISO 31000:2009E.

COSO. (2004). *Enterprise risk management—Integrated framework.* Executive Summary Framework.

Dickinson, G. (2001). Enterprise risk management: Its origins and conceptual foundation. *The Geneva Papers on Risk and Insurance*, *26*(23), 360–366. http://www.jstor.org/stable/41952578

Fell, R., Ho, K. K. S., Lacasse, S., & Leroi, E. (2005). A framework for landslide risk assessment and management. In O. Hungr, R. Fell, R. Couture, & E. Eberhardt (Eds.), *Landslide risk management* (pp. 3–26). Taylor & Francis.

Klinke, A., & Renn, O. (2001). Precautionary principle and discursive strategies: Classifying and managing risks. *Journal of Risk Research*, *4*(2), 159–173. https://doi.org/10.1080/136698701750128105

Klinke, A., & Renn, O. (2004). Systemic risks: A new challenge for risk management. *EMBO Reports*, *5*(1), 41–46. https://doi.org/10.1038/sj.embor.7400227

Konstantinovich, C. H., Cholu-ogly, A. K., & Mikhailovna, C. M. (2019, November 14). Marketing risks of the digital economy. In *International conference of process management and scientific developments, United Kingdom.*

Kotler, P., & Armstrong, G. (2016). *Principles of marketing* (6th ed.). Pearson Education.

Kotler, P., & Keller, K. L. (2016). *Marketing management.* 15th Edition, Pearson Education Limited.

Kuznik, T. (2016). Risk management in a VUCA world: Practical guidelines based on the example of a multinational retail group. In O. Mack, A. Khare, A. Kramer, & T. Burgartz (Eds.), *Managing in VUCA world* (pp. 77–99). Springer International Publishing.

Liuzzo, G., Bentley, S., Giacometti, F., Bonfante, E., & Serraino, A. (2014). The term risk: Etymology, legal definition and various traits. *Italian Journal of Food Safety*, *3*(2269), 36–39. https://doi.org/10.4081/ijfs.2014.2269

Markowitz, H. M. (1991). Foundations of portfolio theory. *The Journal of Finance, 46*(2), 469–477. https://doi.org/10.2307/2328831

Office of Chief Risk Officer, The. (2023). https://ocro.stanford.edu/erm/key-definitions/definition-risk

Outreville, J. F. (1998). *Theory and practice of insurance.* Springer Science+ Business Media.

Power, M. (2004). The risk management of everything. *The Journal of Risk Finance, 5*(3), 58–65. https://doi.org/10.1108/eb023001

Renn, O. (1992). Concepts of risk: A classification. In K. Sheldon (Ed.), *Social theories of risk* (pp. 53–79). Praeger.

Renn, O. (2008). *Risk governance: Coping with uncertainty in a complex world* (1st ed.). Routledge.

Zio, E. (2007). *An introduction to the basics of reliability and risk analysis.* World Scientific Publishing.

Part 4: Leading Digital Transformation in a VUCA World

Dr. Berkan Çelik

Being on the Same Ship with Stakeholders: Industry 5.0

Abstract
Digital technologies are changing rapidly and parallel to this, the reflection of these changes is tried to be moved to the education field as well. However, it is not easy to directly observe these changes in education. This chapter focuses on developing new-age educational leaders and leadership with respect to Industry 5.0 in order to help facilitate these changes in educational environments. This chapter will start by providing brief information about the previous industrial era and then continue with explaining the underpinnings and components of Industry 5.0, its sustainable, human-centric, and resilient nature. Although Industry 5.0 is surrounded by a fast process of technology and digital transformation, the human touch is an undeniable requirement during this process and transformation. Therefore, history of industrial revolutions, reflections on daily life, new-age leadership and digital transformation, and emergence of new professions and skills will be the main topic of interest in this chapter. The chapter will end by recommending how we can make best use of Industry 5.0 and how we can develop new-age educational leaders.

Keywords: Industry 5.0, new-age leadership, digital transformation, digital leaders

The impact of digitalization and its consequences have been significant so far, and they are expected to continue to have important consequences for society in the future (Hensellek, 2020). Technological advancements are a social phenomenon that alters almost every aspect of society (Mokyr, 2000), including education. Compared to past industrial revolutions, today's transitions occur more rapidly and have a profound impact on our way of living, similar to how our ancestors transitioned from hunting and gathering to an agricultural way of life based on irrigation, and later, from an agricultural to an industrial civilization (Rifkin, 2011). In present times, the current commercial landscape is marked by complexity, dynamism, intense rivalry, rapid technological advancements, and unpredictable market demands (Elnadi & Abdallah, 2023). Similarly, "disruption, innovation, turbulence, change, and competition" are the terms that describe the contemporary world we live in (Brett, 2019, p. 3). History has shown that civilizations face critical moments where they must adapt to a new future or face the prospect of decline (Rifkin, 2011). The interactions between industry and society have become more deeply intertwined since the first industrial revolution (Huang et al., 2022). The past decade has seen numerous innovative technologies and applications emerge, resulting in significant changes across industries (Sigov et al., 2022) as well as societies. Significant changes in most industries have created new sectors such as data science, new business models like platforms, new types of companies like cloud computing providers, and new organizational roles like social media account managers (Karacay, 2018). The business world is currently more competitive and fast-paced, with technological innovation and uncertain market demands (Elnadi & Abdallah, 2023). With each new industrialization, there are opportunities to discover and challenges to overcome by optimizing rules and guidelines. Securing a job opportunity in these new industries may be difficult due to the higher knowledge requirements compared to previous industrial revolutions. Despite the potential consequences, industrial revolutions must occur as the world continues to evolve rapidly (ElFar et al., 2021).

Despite the fact that the world is still attempting to adjust and fully comprehend the capabilities of Industry 4.0, certain industrialists and academics have begun to envision and debate the following Industrial Revolution, which is referred to as Industry 5.0 (Di Nardo & Yu, 2021; Raja Santhi & Muthuswamy, 2023). The current understanding of Industry 5.0 is still in its early development stage (Ghobakhloo et al., 2022). Industry 5.0 is a relatively new field of research. The first article related to it appeared in 2016 in Scopus, and the significant increase in publication occurred in 2019 (Madsen & Berg, 2021). Being not a replacement for Industry 4.0, Industry 5.0 is a sustainable future industry that moves beyond traditional productivity and consumption-driven economic models. It has the potential to address various economic and socio-environmental issues, such as labor market disruption, environmental degradation, income polarization, and inequitable regional development (Ghobakhloo et al., 2022). Industry 5.0 is seen as a coherent vision for the future of European industry as well (Breque et al., 2021). Although the Industry 4.0 concept has not yet gained momentum in several developed and developing countries, several sectors that are strongly linked to digital technologies have proposed various conceptual interpretations and visions of the upcoming Fifth Industrial Revolution (Raja Santhi & Muthuswamy, 2023). Some proposed definitions can be exemplified as follows:

- Industry 5.0 is merging the creativity and skill of human beings with the efficiency, productivity, and consistency of robots. It seeks to foster greater appreciation for collaboration between humans and robots, combining their unique strengths to create a more inclusive and human-centered future (European Economic and Social Committee, 2018).

- Industry 5.0 is a new era in the industrial sector, involving a harmonious collaboration between humans and machines, where technology is designed to augment human capabilities in a user-friendly manner (Longo et al., 2020).
- The vision of Industry 5.0 is to achieve societal objectives that extend beyond employment and economic expansion, by ensuring that production adheres to the limits of our planet and prioritizes the well-being of workers in the manufacturing process. It envisions industry as a durable source of prosperity that is capable of withstanding challenges and maintaining sustainability (Breque et al., 2021).
- According to Japan, Industry 5.0 is referred to as "Society 5.0," representing a revolution that focuses on the human touch. This type of society aims to balance economic progress with the resolution of social issues by creating a system that extensively integrates the internet and physical space (ElFar et al., 2021).
- Industry 5.0 is the collaborative efforts of humans, machines, and robots working together to increase productivity in a sustainable manner (Tiwari et al., 2022).

The definitions support that Industry 5.0 is characterized as a cutting-edge manufacturing model that emphasizes the collaboration between humans and machines. This approach aims to boost production efficiency by combining the power of advanced machinery with the innovative potential of human beings (Adel, 2022). With the advent of Industry 5.0, debates now center around fundamental questions regarding humanity's existence, including our physical well-being and relationship with the environment. Although currently a theoretical discussion, it is predicted to become increasingly important in the near future (Sachsenmeier, 2016).

The Fifth Industrial Revolution aims to enhance process efficiency by combining the intelligence of humans and machines on the factory floors. Unlike Industry 4.0, which prioritizes automation, Industry 5.0 focuses on pairing human workers with autonomous machines to leverage human creativity and brainpower. By integrating workflows with intelligent systems, Industry 5.0 seeks to achieve a synergistic collaboration between humans and machines (Nahavandi, 2019). While it is wise to expect new revolutions from Industry 5.0, some support the idea that it is more appropriate to refer to the upcoming revolution as "Industry 4.0S" or "Sustainable Industry 4.0" rather than Industry 5.0 because of the fact that it is still premature to label the upcoming changes as Industry 5.0

without fully comprehending what the future holds in a few decades (Raja Santhi & Muthuswamy, 2023). In order to equip students for success in the Fourth or even Fifth Industrial Revolution, there needs to be a significant shift in the concept of education. The various changes we are witnessing have highlighted the fact that schools must not only prepare children for specific roles, but also for a broad range of opportunities. Simply following a traditional schooling approach is no longer sufficient. What worked for adults in the past may not work effectively for students today and in the future (Sheninger, 2019). This is also what makes educational leaders very important in this era. Before talking about Industry 5.0 and new-age leadership, it is useful to remember the processes that brought us here briefly.

History of Industrial Revolutions

The Industrial Revolution originated in Britain and had a significant impact on global history, particularly in Europe, Asia, and the United States, ultimately contributing to the creation of the modern world (Loy et al., 2021). The Industrial Revolution marked a shift toward industrial and transportation improvements, where handcrafted goods were replaced by the use of advanced technology such as computers and machines to produce goods on a larger scale. This shift in focus has led to the emergence of modernized manufacturing facilities that rely on the most efficient and effective techniques (ElFar et al., 2021). The history of manufacturing has evolved through different industrial revolutions. There have been four distinct revolutions, starting from the late 1700s to the present day. During the First Industrial Revolution, animals, water, and steam engines were used to assist human labor on machines. In the Second Industrial Revolution, electricity and steel were utilized in factories for mass manufacturing. The Third Industrial Revolution saw the incorporation of computer technology into productions and a shift to digital technologies. The current Fourth Industrial Revolution emphasizes the use of digital innovation, IoT (Internet of Things), real-time data, and cyber systems in every sector (Khalid et al., 2021). In general, urban areas experienced positive effects from the Industrial Revolution, including increased wealth, greater quantity and quality of products, and an improvement in living standards. As a result, people began to recognize the significance of education and started attending school classes more regularly (Loy et al., 2021).

First and Second Industrial Revolution (Industry 1.0 and Industry 2.0)

The Industrial Revolution was a period of significant change and innovation that had a profound impact on the world's development. The First Industrial Revolution, which took place from 1760 to the 1830s, marked a shift toward manufacturing in Europe and the United States, while the second, known as the Technological Revolution, began around 1870 and continued until the 1914s. The focus during this period shifted from major inventions to smaller, more incremental improvements aimed at increasing productivity and improving the quality of products. The Second Industrial Revolution, or Industry 2.0, took place in the United States and relied on electricity and crude oil as its primary energy sources. It saw the emergence of the telephone and the widespread use of automobiles for transportation. These two revolutions, with their advancements in mechanization, steam power, weaving looms, mass production, assembly lines, and electrical energy, greatly benefited humanity. They transformed Europe and later spread to Asia and the United States, leading to the formation of the modern world (Leong et al., 2021; Loy et al., 2021).

Third Industrial Revolution (Industry 3.0)

During the 1960s, Industry 3.0 was introduced and started to gain popularity in society. During this time, the use of robots in manufacturing has increased significantly. The primary reason for the proliferation of industrial robots is attributed to the desire to eliminate or minimize the dull, dangerous, and dirty jobs. However, other important factors driving the adoption of robots include the need for consistent quality and flow in manufacturing (Østergaard, 2018). The convergence of internet-based communication technology and renewable energy sources increased the rate of the Third Industrial Revolution. This revolution was expected to have a significant impact on society in the twenty-first century, comparable to the changes brought about by the First Industrial Revolution in the nineteenth century and the Second Industrial Revolution in the twentieth century. In the long run, as with these previous revolutions, the Third Industrial Revolution brought about profound changes to every aspect of our work and daily lives (Rifkin, 2011). The progression from Industry 3.0 to Industry 4.0 has brought about numerous advantages and ease of use as a result of advancements in automation and information technology. The digital network has enabled people and industries worldwide to connect, exchange and share ideas and technological knowledge (Tan, 2021).

Fourth Industrial Revolution (Industry 4.0)

In 2011, Germany introduced a proposal for a new high-tech economic policy concept called Industry 4.0 (Mosconi, 2015). Following the occurrence of the Hannover Fair in Germany in 2013, the term "Industry 4.0" has gained popularity in the manufacturing sector (Lee & Kao, 2013). Industry 4.0 is an advancement from Industry 3.0, which emphasizes standard operations, whereas data analytics originates from statistics and computer science, which prioritize intelligent decision-making (Duan & Da Xu, 2021). The term Industry 4.0 describes the new production processes that rely heavily on technology and automated devices that communicate independently along the value chain activities (Karacay, 2018). When Industry 4.0 was initially introduced, it consisted of nine pillars which included "cyber-physical systems, the Internet of Things, Big data, 3D printing, robotics, simulation, augmented reality, cloud computing, and cyber security" (Yang & Gu, 2021, p. 1312).

The idea of Industry 4.0, with its goal of almost complete automation and the potential cost savings that come with it, has become very appealing to the industry (Østergaard, 2018). Apart from manufacturing, other sectors such as agriculture, energy, healthcare, and many more also stand to gain significant advantages from Industry 4.0 (Duan & Da Xu, 2021). Modern technology plays a vital enabler role in implementing Industry 4.0 successfully, including IoT, Cyber-Physical Systems (CPS), and cloud computing (Elnadi & Abdallah, 2023). The key enablers driving the Fourth Industrial Revolution can be identified as ubiquitous accessibility of the internet and IoT, convergence of business and manufacturing, creation of virtual replicas of real-life scenarios, and implementation of optimized manufacturing processes and smart products (Nahavandi, 2019). Companies aim to achieve several anticipated advantages by implementing these technologies, such as enhancing product quality, refining the decision-making process, cutting down operational expenses, and augmenting productivity levels (Gallab et al., 2021). Briefly, the adoption of Industry 4.0 is anticipated to enhance manufacturing companies' operational efficiency and productivity, flexibility, decision-making, and ability to meet customers' demands, resulting in increased revenue (Elnadi & Abdallah, 2023).

The current revolution has access to a much more limited range of technology than previous industrial periods, but still provides a large boost to businesses, and these technologies will continue to be a prominent feature in Industry 5.0 (ElFar et al., 2021). There are some companies and industries that have not yet embraced Industry 4.0 and digital technologies. This is concerning as these technologies will lay the foundation for the Fifth Industrial Revolution. It is expected that the next revolution will bring changes to business and management practices, particularly in terms of collaboration between humans and smart technologies (Paschek et al., 2019). Since its introduction in 2011, Industry 4.0 has had a significant impact on nearly all industries up to the present time (Yang & Gu, 2021). In brief, the Fourth Industrial Revolution, also known as Industry 4.0, is an advancement in technology that improves productivity. Industry 4.0 has played a significant role in the rapid modernization of the world, especially in the manufacturing industry, and has transformed technology and the way people live their lives (Khalid et al., 2021). As a result of the growing emphasis on human/user-centered products and increased human-machine interaction, it is believed that Industry 4.0 is gradually transitioning toward Industry 5.0 (Aslam et al., 2020).

Fifth Industrial Revolution (Industry 5.0)

Industry 5.0 builds on the paradigm of Industry 4.0 and aims to create a sustainable, human-centered, and resilient industry. It emphasizes the importance of research and innovation in driving this transition and shifts the focus from maximizing shareholder value to considering the interests of all stakeholders involved (Breque et al., 2021). Industry 5.0 aims to balance economic growth and socio-environmental development through the integrative collaboration of industry and stakeholders. Stakeholder-centricity is the major point of differentiation that allows Industry 5.0 to address the socio-environmental shortcomings of Industry 4.0 (Ghobakhloo et al., 2022). Throughout previous industrial revolutions, efforts were made to eliminate or distance humans who were deemed less productive and efficient compared to automation and digital technologies. However, for the first time, the Fifth Industrial Revolution has recognized the importance of incorporating humans and promoting sustainability as additional fundamental principles (Raja Santhi & Muthuswamy, 2023).

Industry 5.0 focuses on three areas of development: human-centric, sustainable, and resilient. Unlike Industry 4.0, which aimed to eliminate humans from the production process, Industry 5.0 acknowledges the power of technology for business development while combining social and ecological responsibility. The symbiosis of technological, social, and ecological segments constitutes the essence of Industry 5.0 (Grabowska et al., 2022). Industry 5.0 aims to place human beings once again at the forefront of industrial production, using tools like collaborative robots. This approach not only satisfies current consumer demands, but also provides workers with more fulfilling jobs than those typically associated with factory work for the past hundred years or so (Østergaard, 2018).

Industry 5.0 is advancing across various sectors including retail, healthcare, finance, manufacturing, textile, education, food, and more (Adel, 2022).

Industry 5.0 is characterized by the widespread use of data and knowledge from society, which distinguishes it from previous industrial periods (ElFar et al., 2021). Prospective Industry 5.0 technologies (Adel, 2022; Raja Santhi & Muthuswamy, 2023; Tiwari et al., 2022) are:

- Collaborative robots,
- Smart sensors,
- Digital twins,
- Internet of Things (IoT), Internet of Everything (IoE) and Artificial Intelligence of Things (AIoT),
- Blockchain,
- Edge and fog computing,
- Cognitive computing,
- 6G systems and beyond,
- Augmented reality (AR), mixed reality (MR) and holography,
- Big data and big data analytics,
- Artificial intelligence (AI),
- Multi-agent systems and technologies,
- Smart manufacturing,
- Smart materials,
- Adaptive systems,
- Smart materials,
- 3D, 4D, and 5D printing,
- Digital ecosystem.

While some of the technologies mentioned in the Industry 5.0 enablers may appear to be only slight improvements or upgrades to those found in Industry 4.0, a closer examination of each technology would demonstrate their significant emphasis on the incorporation of human, environmental, and social awareness (Raja Santhi & Muthuswamy, 2023). However, still there is no clear plan for how to use new technologies to initiate the transformation (Salkin et al., 2018).

The concept of Industry 5.0 aims to achieve a symbiotic relationship between humans and machines by augmenting human capabilities with technology as part of a cognitive process (Longo et al., 2020). This is depicted in Fig. 1 in detail.

It is highly likely that the next wave of industrial development, Industry 5.0, will involve a necessary but gradual transformation within the industry. AI will become more prevalent in many organizations and companies, resulting in some jobs being replaced by machines. IoT will have a greater impact on people's daily

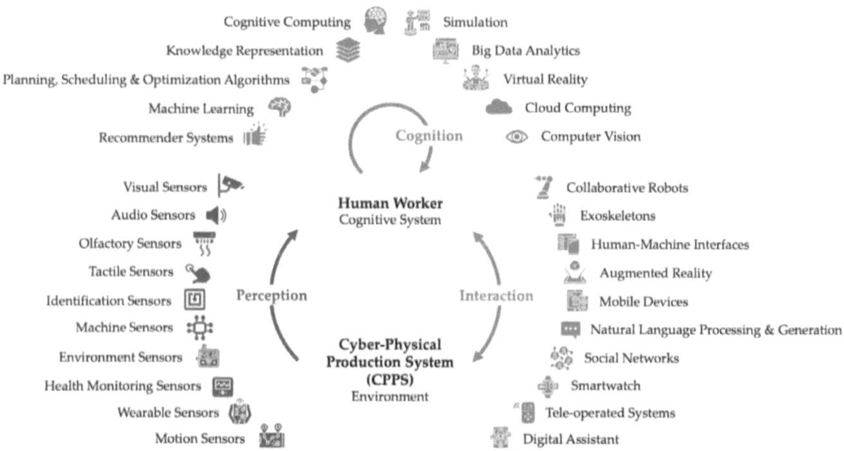

Fig. 1. Relationship between humans and machines by augmenting human capabilities with technology as part of a cognitive process (Longo et al., 2020, p. 4) (Open Access Figure)

lives as their devices become more digitized and interconnected globally. In addition to the societal and environmental benefits of Industry 5.0, renewable energy sources such as wind, solar and biofuels will replace fossil fuels that produce greenhouse gasses, thereby improving the environment. While the specifics of how these technologies will be implemented are not certain, their benefits are undeniable (Leong et al., 2021). The concept of Industry 5.0 represents a shift from a technology-driven focus to one that prioritizes societal and human concerns. This change in perspective has the potential to greatly impact a wide range of industries through significant paradigm shifts in the near future (Jafari et al., 2022).

Reflections on Daily Life

Predicting the future path of Industry 5.0 is a challenging task (Madsen & Berg, 2021). The questions have aroused drawing from previous experiences of industrial revolutions, will we be better equipped to manage the transition to Industry 5.0, or will it remain a disruptive, harsh, and socially turbulent shift from one industrial era to the next? (Sachsenmeier, 2016). More questions arise with Industry 5.0 as follows: In what ways will Industry 5.0 impact the responsibilities of workers, as well as the organization of work? What new skills must be acquired as we work alongside industrial robots? How can we fully utilize human intelligence, creativity, and dexterity to our advantage? Moreover, as we approach the next stage of the industrial revolution, it is essential to establish guidelines for human-machine interaction and the involvement of AI. It is crucial to consider the potential conflicts that may arise between humans and AI (European Economic and Social Committee, 2018). In the near future, it may be necessary to provide ethics training to executives, managers, and employees. Additionally, there may be a need for a 24-hour hotline that offers ethical guidance in situations where ambiguity arises, which could benefit both workers and employers (Longo et al., 2020).

As anticipated, discussions have emerged regarding the societal implications of Industry 4.0 and related concepts, particularly with regards to employment (Sachsenmeier, 2016). Missing digital competence, missing technology and infrastructure, missing skills and need for guidance for transformation are the main concerns for transition to Industry 5.0. As a result, questions are raised about what skills need to be developed and acquired, what rules should govern human-machine interaction, and what effects and conflicts may arise from the integration of AI (Paschek et al., 2019). Ensuring that technology suits the needs of humans involves educating those who will use it. At present, the primary obstacles include security and privacy concerns, a shortage of skilled workers, as well as the time-consuming and costly nature of implementing new technology (Adel, 2022). Consequently, the demand for highly skilled employees will increase, posing a challenge for organizations to meet this need (Sony, 2020).

The emphasis on technological advancements in Industry 4.0 has largely disregarded the importance of human intelligence and overlooked the potential negative impacts of digital technologies on the environment and society. Similar to previous industrial revolutions, Industry 4.0 technologies have the capability to contribute to automation and disconnection of humans from the factory shop floor, resulting in job losses and social inequality. Therefore, it is crucial to either avoid the use of these technologies or upgrade them to prioritize the inclusion of humans in the system and focus on eco-friendliness. The upcoming Industrial Revolution is expected to fulfill these needs (Raja Santhi & Muthuswamy, 2023). As the Fifth Industrial Revolution is purposely structured, it highly values the advancement of the workforce. New competencies, qualifications, and even careers such as "CRO (Chief Robotics Officer), device instructor, solution coach, process supervisor, robot aide, machine coder, mobile robot administrator, cyber-physical systems analyst, collaborative operator, device connector, artificial intelligence operator, virtual operator" are arising (Grabowska et al., 2022, p. 3142). Thus far, the business models have not been significantly affected by these concepts, and they require further development before they can be implemented as industrial realities (Sachsenmeier, 2016). In general, gaining further knowledge and understanding of collaborative robotics and the interaction between humans and machines are essential (European Economic and Social Committee, 2018). A new era of industrial development, known as the Fifth Industrial Revolution, will come to fruition when three pivotal components—smart devices, advanced systems, and automated processes—seamlessly integrate with the tangible world, working alongside human intelligence (Nahavandi, 2019). These emphasize the requirement to prioritize the establishment of a suitable education and training framework for individuals supported by educational leaders so that we can be ready for the upcoming revolution. After the inception of Industry 4.0, the terms such as Society 4.0 or Education 4.0 have appeared on the stage. The same trend is valid for Industry 5.0 as well, which is followed by Society 5.0 or Education 5.0. Production environments have become increasingly complex, making conventional training methods obsolete. For the requirements of Education 5.0, there is a need to incorporate modeling & simulation techniques with virtual reality to provide a useful training environment (Longo et al., 2020). It is essential to incorporate technology-based educational models such as MOOCs into the education approach (Kulal & Dinesh, 2023) at all levels. There is also a need to implement new and innovative education systems, but this will not necessarily solve the problem for older workers. This is a complex issue that may require significant time and effort to resolve (Sung, 2018).

New-Age Leadership and Digital Transformation

The Education 4.0 era has led to a growing need for school reforms, which, in turn, are challenging the traditional roles of school leaders (Kin & Kareem, 2019). The education system is evolving, but leaders often adhere to outdated norms and maintain a framework from a previous era, which stifles experimentation and the potential scalability of digital innovation (Kowch, 2021). Initially, early studies on leadership were focused on determining the characteristics and behaviors of good leaders in order to identify potential leaders who could succeed in their roles. However, more recent trends in the leadership literature indicate a shift toward a more practical question for the present times. Specifically, the focus is now on how to aid individuals in becoming and continuing to be effective leaders. This new approach acknowledges that the match between individual and their jobs is not fixed and that there is no one perfect state of leadership effectiveness. Instead, individuals must continuously develop and evolve to be successful leaders (Dai & De Meuse, 2021). Educational sector requires individuals to take on various leadership roles (Arham et al., 2022). The modern leader in the era of digital transformation is significantly different from a traditional leader. Simply receiving university training is not enough to develop leaders who meet the requirements of the digital age. Effective mentoring and modern development programs are also necessary for new leaders to develop new ways of thinking, acting, and reacting (Gubernatorov et al., 2021). The digitization of work has increased the importance of leadership (Larson & DeChurch, 2020). There is not universally one best leadership style for every situation (Oberer & Erkollar, 2018). However, the fundamental essence of leadership, which is becoming more and more human-centered, will continue to be vital and significant (Jakubik & Berazhny, 2017). A new leadership paradigm seems to be emerging due to the current digital disruption we are facing (Tigre et al., 2023). It is equally important to adopt appropriate leadership behaviors for this era. Consequently, leaders must abandon their traditional leadership styles and embrace contemporary leadership styles that are necessary for this new age (Cinnioğlu, 2020). In the digital age, leadership is becoming increasingly important and essential. The leaders of today's digital era must acquire new skills, abilities, and adopt a different mindset. Conventional leadership models may not suffice to equip future leaders with the necessary competencies (Maheshwari & Yadav, 2020). In all sectors, including developed and emerging economies, digital leadership is a critical concern for organizations worldwide (El Sawy et al., 2016).

The need for further digital transformation efforts is evident to adapt to the new business environment (Elnadi & Abdallah, 2023). However, the success of new technologies relies heavily on people's trust and acceptance of them, and therefore, it is essential to train individuals to effectively use these tools as technology becomes more integrated with human tasks (Adel, 2022). To realize this, new-age leaders have a greater role and responsibility. One critical skill that leaders must possess is leadership agility, which refers to their capacity to adapt to changes effectively and promptly in the business environment by taking targeted and flexible actions. By equipping their team for a volatile, uncertain, complex, and ambiguous (VUCA) world, leaders can enable employees to acknowledge and proactively respond to impending changes, instead of being unaware of potential changes that could affect the organization (Horney et al., 2010). Leaders who possess the capacity to think in a flexible manner when confronted with changing environmental conditions can effectively adapt their organization to these alterations (Akkaya & Sever, 2022). Thus, in order to successfully implement digital transformation efforts and navigate the evolving business landscape, leadership agility is a critical skill that leaders must possess (Akkaya & Sever, 2022; Horney et al., 2010).

Digital Leadership

The twenty-first century is witnessing a surge of innovation fueled by the ongoing digital revolution and the growth of worldwide markets. The present circumstances of rapid uncertainty and transformation are unlikely to diminish and stabilize anytime soon (Horney et al., 2010). Currently, we are confronted with the task of not only acknowledging the emergence of new technologies but also integrating them to achieve success (Sigov et al., 2022). Globalization, advancements in communication and information technologies, and heightened competition have caused conventional leadership styles to be substituted with contemporary ones. Nonetheless, these factors are persistently evolving and transforming. As a result, a challenge arises in determining the specific types of leadership that are suitable for the modern era (Cinnioğlu, 2020).

Due to the increasing complexity of modern working conditions, traditional leadership and management approaches are no longer effective. In order to effectively manage and lead in today's work environment, it is important to be highly skilled and effectively handle complex data. Therefore, leadership and management models which embrace the VUCA concept should be adopted (Akkaya & Sever, 2022). To implement and adopt Industry 5.0 in educational contexts, educational leadership comes to the fore. Educational leadership in the form of digital leaders with varying properties can be helpful for the transition to Industry 5.0. Digital leadership can be the possible new-age leadership in the digital realm as digital leaders can be strong to meet the challenges of digitalization as well as actively shape it. It is important to recognize that digital leadership is not a fixed concept, and it will inevitably develop and change as society continues to embrace digitalization (Hensellek, 2020). According to Arham et al. (2022, p. 2769), digital leadership can be defined as "the ability to lead the people and organization toward digital transformation and being agile in a rapidly changing digital landscape." Reflecting the contrast between leadership and management, digital leadership can also be defined as "doing the right things for the strategic success of digitalization for the enterprise and its business ecosystem" (El Sawy et al., 2016, p. 142). Digital leadership is not limited to top management positions and cannot be approached solely from a top-down perspective. A more comprehensive approach is required, and employees at all levels of the hierarchy should be encouraged to participate in digital leadership (Hensellek, 2020). However, a standard model or a shared comprehension of digital leadership is absent (Klein, 2020). Digital leadership involves a rapid and collaborative approach that transcends hierarchical boundaries and emphasizes innovation (Oberer & Erkollar, 2018). Leaders who hold formal positions can hinder interaction and innovation by isolating individuals from the challenges and opportunities that could drive change in schools. Similarly, formal organizations can restrict the room for digital innovations by imposing regulations and vertical functional hierarchies that impede collaboration among individuals possessing diverse knowledge and skills across the school ecosystem (Kowch, 2021). The importance of digital leadership is increasing as it brings about several benefits. Teachers with higher levels of ICT-supportive leaders are more likely to report higher levels of perceived usefulness of computer technology, perceived learning outcomes for students and more frequent use of computer technology compared with teachers reporting lower levels of ICT-supportive leaders (Edvard Hatlevik & Christian Arnseth, 2012). Digital leadership is also closely associated with digital transformation (Tigre et al., 2023), and the effective use of technology in schools is enhanced by the digital leadership of principals (AlAjmi, 2022). If school principals exhibit their roles as digital leaders and promote the corporate digital learning culture, it can greatly aid in achieving digital transformation in education and promote a digital learning culture within schools (Karakose et al., 2021).

According to an analysis of existing leadership concepts, both transformational and entrepreneurial leaders create a future vision and pursue it. This also seems especially important in the context of digitalization since the technology presents opportunities that are still challenging for many people to comprehend (Hensellek, 2020). As leaders adopt mindsets that acknowledge the complex and adaptive nature of educational ecosystems, they can develop the necessary skills (Kowch, 2021). Furthermore, leadership agility refers to a leader's capacity to adapt to changes effectively and promptly in the business environment by taking targeted and flexible actions. This skill allows the leader to equip their team for a VUCA world by enabling them to adjust their mindset and relevant skills. Instead of being unaware of potential changes that could affect the organization, employees are empowered to acknowledge and proactively respond to impending changes (Horney et al., 2010). Overall, to survive in today's VUCA environment, educational leaders should be responsible visionary agile transformational digital leaders.

Characteristics of Digital Leaders

Educational leaders', especially digital leaders' success in the transition to Industry 5.0 and also in the Industry 4.0 era is contingent upon possessing certain characteristics. Although certain fundamental leadership skills stay consistent, the distinct features of digital transformation necessitate the acquisition of new capabilities (Tigre et al., 2023). As schools adopt programs aimed at improving their effectiveness, school leaders must develop the necessary competencies to bring about an effective transformation of the education system (Kin & Kareem, 2019). For successful digital transformation in K-12 education, school principals must exhibit their digital leadership skills and encourage the formation of a digital learning culture in their schools (Karakose et al., 2021). Generally, to thrive in today's fast-changing landscape, organizations require leaders with a unique set of skills. These leaders should possess a forward-thinking approach, the ability to identify technological advancements, and quickly change direction when required. Moreover, it is vital for leaders to be more agile, enabling them to respond rapidly to potential risks and opportunities (Friedman, 2020). The effectiveness of a 4.0 leader is influenced by their personal competencies, including their mindset and proficiency in using new methods and tools such as design thinking. Design thinking is a problem-solving approach utilized by designers to develop desirable solutions, and a design mindset is a proactive and outcome-driven attitude toward building a better future (Oberer & Erkollar, 2018).

Several research studies have focused on exploring or revealing the characteristics of digital leaders using empirical research or systematic reviews. In a study examining teachers' perspectives on school principals' digital leadership roles and technology capabilities, school principals' digital leadership skills were categorized into three groups: technology use, management skills, and personal skills (Karakose et al., 2021, p. 13):

- *Technology usage:* Technology capabilities, encouraging digital technology, keeping up with technology, building a digital school culture, digital literacy.
- *Managerial skills:* Change management, innovation management, cooperation, visionary, entrepreneurship, treating everyone equally, risk management, sharing responsibility, trust, participation in decisions, motivation, accountability, managing individual differences.
- *Personal skills:* Being open to learning, communication skills, creativity, critical thinking, problem-solving ability, practical intelligence, global thinking (Karakose et al., 2021, p. 13).

According to Tigre et al.'s (2023, p. 57) bibliometric analysis on digital leadership, integrated elements for the digital leaders' main capabilities were found as follows: People focus such as communication, trust-based relationship, Personal aspects such as ethics and adaptability, Long-term orientation such as innovation, vision, and direction, and Task achievement such as digital savvy and collaborative work. A digital-savvy leader demonstrates a preference for interpersonal skills, including successful collaboration with others and personal traits such as emotional management. Additionally, they possess strategic vision to guide the organization toward future objectives and are proficient in delivering intended outcomes. The new digital leadership model appears to possess leadership qualities that are necessary for success in a digital setting. These traits include effective communication, providing direction, transparency, building trust, agility, collaboration, promoting innovation, empowerment, and being adaptable to changing circumstances (Tigre et al., 2023).

Hensellek (2020) suggested a potential framework to practitioners and researchers that can help them overcome the obstacles of digitalization and even take an active role in shaping it. The framework for digital leadership includes three essential components—a digital vision, digital mindset, and digital skillset—that digital leaders should have or acquire. While having the right attitude is essential, a digital leader should possess a digital skillset which includes the required abilities to comprehend digital technologies, use them proficiently, and make sensible decisions. It is crucial for leaders to identify opportunities and evaluate risks associated with digitalization, which is closely tied to a digital mindset and is a continuous learning process. Meeting all three requirements is essential to be considered a digital leader and enables them to formulate and achieve their digital vision (Hensellek, 2020). Previous leadership styles like transformational or entrepreneurial leadership also provide an important foundation for the digital leadership framework introduced here, which make them useful. This digital leadership framework can be applicable for educational leaders as well as given in Fig. 2.

What Should Digital Leaders Do to Be Ready for the Industry 5.0 Era?

Digital transformation competency is among the enablers of Industry 5.0, which are crucial to achieving sustainable industrial transformation objectives (Ghobakhloo et al., 2022). To establish quality and standards in education, transformation is necessary. This involves transforming the roles and skills of students, teachers, and headmasters to adapt to digital technologies and technology-supported learning environments, as education for the future requires. Being digital leaders for future education is therefore essential for quality and innovation (Aksal, 2015). Industry 5.0 actors must develop and implement new approaches that address social welfare issues, such as the digital divide, and skill gap. This requires organizational changes (Ghobakhloo et al., 2022). Concerning the change and leadership, new-age educational leaders should perform modifications in organizational culture and supervise and regulate them. They should consider potential resistance and reluctance to change. It is vital, from a practical

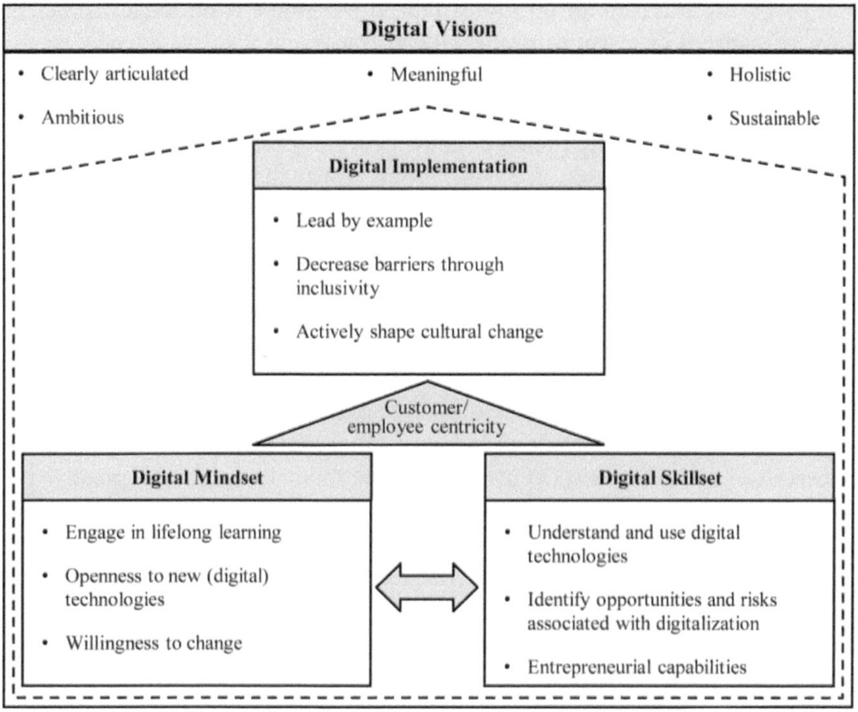

Fig. 2. The digital leadership framework by Hensellek (2019) (Hensellek, 2020, p. 63) (Open Access Figure)

standpoint, to establish particular use cases and cultivate digital competencies (Schneider, 2018). A digital leader must possess a determination to bring about change. They should frequently challenge the existing state of affairs and continuously work toward improvement in order to keep up with the rapid changes in today's fast-paced environment (Hensellek, 2020). Educational policies serve as the primary means of shaping the human resources of the future. Emerging technologies will redefine the labor market as we presently understand it. Given that contemporary students remain connected through advanced technologies and social networks, the education sector must adapt its curricula and strategies to meet the opportunities and challenges presented by this transformation (Cotet et al., 2020). With society experiencing exponential change, it is imperative that education equips learners for a world that is unpredictable and challenging to forecast. All stakeholders in education have a responsibility to acknowledge that preparing students for an industrialized workforce is no longer sufficient to meet society's or learners' needs. Digital leadership involves advocating for change that will convert schools into lively epicenters of learning (Sheninger, 2019). In order to address technological challenges, an educational leader should adapt to evolving learning preferences by incorporating virtual and augmented reality, facilitating teaching, fostering group work, and presenting real-world case studies (Cotet et al., 2020). For the leadership in the age of VUCA, it is also essential to have programs in place that train educational leaders on how to effectively utilize and assess digital tools and teaching methods that cater to individual learning needs and prioritize skill-based education. This ensures that all leaders are equipped with digital skills and have access to high-quality digital learning resources (Mohebi, 2019). Leaders should demand this kind of training for themselves or find ways to access this type of training. Among the critical factors that facilitate the adoption of Industry 4.0, technology-related aspects such as perceived usefulness, ease of use, and compatibility had a notable positive influence on the adoption of Industry 4.0 (Raj & Jeyaraj, 2023). Educational leaders should be aware of these adoption factors and develop themselves accordingly.

In addition, the ongoing digital transformation's impact on the workforce and the extent of change it will bring remain uncertain, raising questions about whether it will create jobs or lead to displacement. To address the potential negative effects of automation, it is necessary to prioritize upskilling and reskilling programs to ensure that the labor force's knowledge, skills, and abilities are developed (Hepaktan & Şimşek, 2022). This need is heightened in the context of Industry 5.0, where the focus on seamless man-machine symbiosis requires new-age leaders to acquire necessary technologies and prioritize their workforce's development (Ghobakhloo et al., 2022). Virtual training programs can help create a skilled workforce while minimizing disruptions and risks, leveraging advances in AI, graphical processing units, big data, and haptic technologies (Nahavandi, 2019). Educational leaders should carefully address the following issues. To further facilitate the development of the labor force and support students, educational leaders must introduce emerging professions in schools at all levels, increase awareness among students, and help them develop their knowledge, skills, and abilities. They should also improve themselves on virtual training options.

Lastly, for action and leadership in the realm of digitalization, educational leaders should be aware of the skills value chain in general because it facilitates improved collaboration within the systems for education and training, and it can systematically provide guidance and some insights that educational leaders can benefit from. The concept of a value chain refers to a series of actions a company performs to provide customers with valuable products or services. The skills value chain, explained in Fig. 3, can provide a competitive edge by aligning workforce talents and skills with technological and productivity opportunities. Improving workforce skills alongside technological advances can reduce technology adoption lead times and lead to higher-value employment (Collier & Shakspeare, 2020) in the future. Nevertheless, successfully implementing this approach will necessitate collaborative leadership and the enthusiastic participation of a diverse group of stakeholders who are committed to a common goal and willing to work together. Regarding educational opportunities, to facilitate the acquisition of skills necessary for the future workforce, it is important to establish credit-bearing flexible modular training and resources that are specifically focused on emerging technologies and cross-sector standards. These modules should be designed to assist individuals in acquiring the new and modified skills that will be required (demanded) in the future, and their achievements should be acknowledged. Educational leaders should work on this as it will be difficult to keep up with the developments with the static curriculum.

Emergence of New Professions

In the past, people used to pursue a career right after graduating from university, and they would stick to it until their retirement (ElFar et al., 2021). However, things are changing very rapidly and significantly in today's societies. While it would be ideal for new employees to possess all the skills required for their job, the reality is that the necessary skills and knowledge change rapidly throughout one's career (Gehrke et al., 2015). Soft skills, such as communication and social skills, teamwork, and self-management skills, have become increasingly important. However, the average factory worker usually does not receive training in these areas as their job duties typically do not necessitate the utilization of such skills (Cotet et al., 2020).

To take full advantage of industrial development and its transformative potential, a workforce with the necessary skills aligned with industry-specific requirements must be prepared. The rapid automation of business processes and the

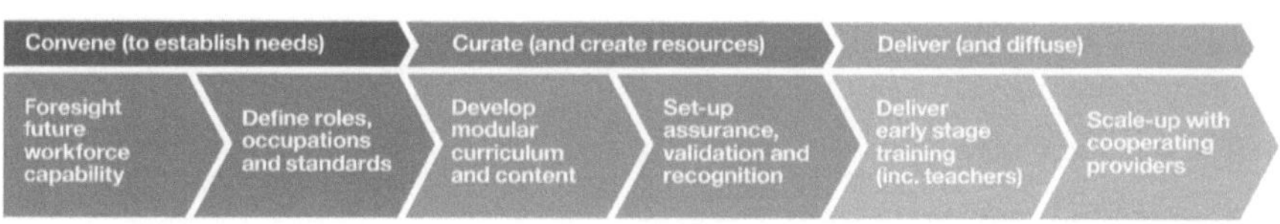

Fig. 3. The skills value chain (Collier & Shakspeare, 2020, p. 9)

emergence of new business models require new skill sets for employees. Further adoption of digital systems and successful implementation of Industry 4.0 will require an even wider range of employee skills due to the increased complexity of work environments and new operational and organizational structures. Therefore, employee roles and work processes will change, and these changes will necessitate significant transformations in job descriptions and skill requirements (Karacay, 2018). Industry 5.0 is expected to bring about significant changes due to the increased selectivity of job vacancies. This trend is likely to lead to an escalation in the employability rate (ElFar et al., 2021). As AI and industrial robots continue to advance, they will replace many of the jobs that exist today. However, this transformation will also create new job opportunities that require vocational skills, creativity, and empathy (Leong et al., 2021). In the future, jobs that are undergoing technological changes will demand both technical expertise and human skills. Even though AI, robots, and smart machines are taking over production, human intelligence remains vital in creating and maintaining these technologies. Even fully automated production and service units require technicians and engineers to ensure their flawless operation. This suggests that while unskilled labor may struggle to compete with technology, there is still a need for qualified labor in this changing job market (Hepaktan & Şimşek, 2022).

According to the estimation based on the analysis of 15 industries and 26 economies, the division of labor between humans and machines may cause 85 million jobs to be replaced by 2025. However, there is a potential for 97 million new jobs to emerge that are better suited to the evolving division of labor between humans, machines, and algorithms (World Economic Forum, 2020). Although there will be job disruptions, there will also be job creation in emerging fields, known as the "jobs of tomorrow." In the next decade, a significant portion of new jobs will be in completely new occupations or existing occupations that undergo significant changes in terms of their required skills and job responsibilities. For those who remain in their roles, it is anticipated that 40% of core skills will change in the next five years, and 50% of all employees will require reskilling. Top 10 jobs with growing job demand and decreasing job demand in various industries are provided in Tab. 1.

Tab. 1. Top 10 Jobs with Growing Job Demand and Decreasing Job Demand (World Economic Forum, 2020, p. 30)

#	Growing Job Demand	Decreasing Job Demand
1.	Data analysts and scientists	Data entry clerks
2.	AI and machine learning specialists	Administrative and executive secretaries
3.	Big Data specialists	Accounting, bookkeeping and payroll clerks
4.	Digital marketing and strategy specialists	Accountants and auditors
5.	Process automation specialists	Assembly and factory workers
6.	Business development professionals	Business services and administration managers
7.	Digital transformation specialists	Client information and customer service workers
8.	Information security analysts	General and operations managers
9.	Software and applications developers	Mechanics and machinery repairers
10.	Internet of Things specialists	Material-recording and stock-keeping clerks

The demand for certain skills is expected to change in the next five years, resulting in continued skills gaps. Employers anticipate that skills such as critical thinking, analysis, problem-solving, self-management, active learning, resilience, stress tolerance, and flexibility will become more important. On average, companies believe that approximately 40% of their workers will need to be reskilled in six months or less. Additionally, 94% of business leaders expect employees to acquire new skills while on the job (World Economic Forum, 2020). Moreover, various technical and personal attributes and proficiencies gain significance in the age of VUCA for the factory of the future (Gehrke et al., 2015, p. 13):

- **Technical qualifications and skills:** IT knowledge and abilities, data and information processing and analytics, statistical knowledge, organizational and processual understanding, and ability to interact with modern interfaces such as human-machine/human-robot,

- **Personal qualifications and skills:** Self and time management, adaptability and ability to change, team working abilities, social skills, communication skills, trust in new technologies, and mindset for continuous improvement (Gehrke et al., 2015, p. 13).

Digital leaders should be ready for the emergence of new jobs and disappearance of others. Furthermore, digital leaders should assist their teachers, students, and stakeholders in being aware of these qualifications and skills, and they should have an active role in creating a vision to help students acquire these skills and qualifications.

Conclusion

While the initial three industrial revolutions unfolded over a period of several decades, changes happen so rapidly in today's society. With the rapid advancement of technology and innovation, there is a potential for revolutions to occur in quick succession over the next decade and beyond (Tiwari et al., 2022). However, the process of digitalization is characterized by continuous change and an uncertain outcome, and therefore, leaders must acknowledge that their work environment and the expectations placed on them, and their staff are evolving (Hensellek, 2020). In Industry 5.0, humans and AI machines will work together more closely and their interactions will become more common and visible in society (Leong et al., 2021). This shift will create new challenges in the area of human-machine interaction (HMI) as machines will become more involved in everyday human life than ever before (Nahavandi, 2019). On the other hand, the advanced technologies of Industry 5.0 have the potential to restore the lost partnership between humans, machines, and software systems, creating a sustainable work environment on the shop floor and promoting sustainability for the entire world (Raja Santhi & Muthuswamy, 2023).

Although technology, manufacturing, and communication industries are seeing positive developments and are beginning to transition toward Industry 5.0, there are still some industries which have not heard about Industry 5.0 yet (Paschek et al., 2019). Parallel to this, it is very likely that these developments can also be difficult to be heard in education as revolutions take place very slowly in education. To ensure successful adoption in education, educational leaders must motivate stakeholders and consider technology-related factors, such as perceived usefulness and compatibility (Raj & Jeyaraj, 2023). Furthermore, digital leaders with a vision for the future and a digital mindset and skillset will be necessary for the execution of the Industry 5.0 transformation (Hensellek,

2020). Development and integration of Industry 5.0 will require proactive governmental support, resource availability and capabilities, digital transformation competency, sustainability orientation and thinking, and stakeholder integration (Ghobakhloo et al., 2022). Digital educational leaders should take their parts in facilitating the transition to Industry 5.0.

In today's VUCA environment, it is essential to have sustainable digital leadership to ensure their long-term success in an increasingly global and digital market (Hensellek, 2020). In order not to lag behind the pace of digitalization, educational leaders should regularly align themselves with the path of digital transformation in Industry 5.0. It is known that the ability to access necessary resources is extremely important in order to develop digitalization capacity, and lack of resources might be a major reason for the digital divide; however, transformation should start with an idea and vision when there are not enough resources, and educational leaders should take an active role in the transformation for Industry 5.0. Exceptional leaders will seek to examine several alternatives and possibilities and consider multiple options and paths as usually, there will not be just one optimal course of action, and there is typically no one best solution in the VUCA world. The VUCA world is characterized by change, which includes both risky ruptures and positive innovation. Even though dangers may be more evident, there are many hidden opportunities that inspiring strategies can be derived from (Johansen, 2007).

To conclude, the growth of industry brings numerous advantages to a community and the local economy, including job opportunities, educational opportunities, and capital efficiency (ElFar et al., 2021). The progress of society is closely linked to the development of industry. The advancement of industry will drive the progress of society, and the evolution of society will facilitate the next Industrial Revolution (Huang et al., 2022). Educational leaders should be aware of the critical factors that facilitate the adoption of Industry 5.0, such as technology-related aspects, and motivate stakeholders to facilitate accentuating Industry 5.0 adoption. Eventually, Industry 5.0 will change qualification requirements completely. For the new-age leadership and digital transformation, educational leaders should ensure that teachers' and students' knowledge and abilities are maintained at a high level. However, it is in new-age leaders' hands that they take part in developing driving enablers to support this transformation.

References

Adel, A. (2022). Future of industry 5.0 in society: Human-centric solutions, challenges and prospective research areas. *Journal of Cloud Computing, 11*(1), 1–15. https://doi.org/10.1186/s13677-022-00314-5

Akkaya, B., & Sever, E. (2022). Agile leadership and organization performance in the perspective of VUCA. In M. R. I. Talukdar, C. Z. Lamagna, C. C. Villanueva, R. Nahar, & F. Hassan (Eds.), *Post-pandemic talent management models in knowledge organizations* (pp. 213–228). IGI Global.

Aksal, F. A. (2015). Are headmasters digital leaders in school culture? *Egitim ve Bilim, 40*(182), 77–86. http://dx.doi.org/10.15390/EB.2015.4534

AlAjmi, M. K. (2022). The impact of digital leadership on teachers' technology integration during the COVID-19 pandemic in Kuwait. *International Journal of Educational Research, 112*, 101928. https://doi.org/10.1016/j.ijer.2022.101928

Arham, A. F., Norizan, N. S., Arham, A. F., Hasbullah, N. N., Malan, I. N. B., & Alwi, S. (2022). Initializing the need for digital leadership: A meta-analysis review on leadership styles in educational sector. *Journal of Positive School Psychology, 6*(8), 2755–2773. https://journalppw.com/index.php/jpsp/article/view/10280

Aslam, F., Aimin, W., Li, M., & Ur Rehman, K. (2020). Innovation in the era of IoT and industry 5.0: Absolute innovation management (AIM) framework. *Information, 11*(2), 124. https://doi.org/10.3390/info11020124

Breque, M., De Nul, L., & Petridis, A. (2021). *Industry 5.0 Towards a sustainable, human-centric and resilient European industry* [Policy brief]. Directorate-General for Research and Innovation. https://research-and-innovation.ec.europa.eu/document/download/2359a188-5702-4831-a529-3dac3f4ab2a3_en

Brett, J. (2019). *Evolving digital leadership: How to be a digital leader in tomorrow's disruptive world.* Apress.

Cinnioğlu, H. (2020). A review of modern leadership styles in perspective of industry 4.01. In B. Akkaya (Ed.), *Agile business leadership methods for industry 4.0* (pp. 1–23). Emerald Publishing. https://doi.org/10.1108/978-1-80043-380-920201002

Collier, I., & Shakspeare, P. (2020). *Manufacturing the future workforce. UK industrial strategy.* https://www.gatsby.org.uk/uploads/education/reports/pdf/manufacturing-the-future-workforce.pdf

Cotet, G. B., Carutasu, N. L., & Chiscop, F. (2020). Industry 4.0 diagnosis from an iMillennial educational perspective. *Education Sciences, 10*(1), 21. https://doi.org/10.3390/educsci10010021

Dai, G., & De Meuse, K. P. (2021). Learning agility and the changing nature of leadership: Implications for theory, research, and practice. In V. S. Harvey &

K. P. De Meuse (Eds.), *The age of agility: Building learning agile leaders and organizations* (pp. 31–61). Oxford University Press. https://doi.org/10.1093/oso/9780190085353.003.0002

Di Nardo, M., & Yu, H. (2021). Special issue "industry 5.0: The prelude to the sixth industrial revolution." *Applied System Innovation*, *4*(3), 45. https://doi.org/10.3390/asi4030045

Duan, L., & Da Xu, L. (2021). Data analytics in industry 4.0: A survey. *Information Systems Frontiers*, 1–17. https://doi.org/10.1007/s10796-021-10190-0

Edvard Hatlevik, O., & Christian Arnseth, H. (2012). ICT, teaching and leadership: How do teachers experience the importance of ICT-supportive school leaders? *Nordic Journal of Digital Literacy*, *7*(1), 55–69. https://doi.org/10.18261/ISSN1891-943X-2012-01-05

ElFar, O. A., Chew, K. W., & Show, P. L. (2021). What is industry 5.0? In P. L. Show, K. W. Chew, & T. C. Ling (Eds.), *The prospect of industry 5.0 in biomanufacturing* (pp. 57–93). CRC Press.

Elnadi, M., & Abdallah, Y. O. (2023). Industry 4.0: Critical investigations and synthesis of key findings. *Management Review Quarterly*, 1–34. https://doi.org/10.1007/s11301-022-00314-4

El Sawy, O., Amsinck, H., Kraemmergaard, P., & Lerbech Vinther, A. (2016). How LEGO built the foundations and enterprise capabilities for digital leadership. *MIS Quarterly Executive*, *15*(2), 5, 141–166.

European Economic and Social Committee. (2018). *Industry 5.0*. https://www.eesc.europa.eu/en/agenda/our-events/events/industry-50

Friedman, H. H. (2020). Organizational agility, visionary leadership in the age of VUCA. *SSRN*. http://dx.doi.org/10.2139/ssrn.3728372

Gallab, M., Bouloiz, H., Kebe, S. A., & Tkiouat, M. (2021). Opportunities and challenges of the Industry 4.0 in industrial companies: A survey on Moroccan firms. *Journal of Industrial and Business Economics*, *48*(3), 413–439. https://doi.org/10.1007/s40812-021-00190-1

Gehrke, L., Kühn, A. T., Rule, D., Moore, P., Bellmann, C., Siemes, S., Dawood, D., Lakshmi, S., Kulik, J., & Standley, M. (2015). A discussion of qualifications and skills in the factory of the future: A German and American perspective. *VDI/ASME Industry*, *4*(1), 1–28. https://www.vdi.de/ueber-uns/presse/publikationen/details/industry-40-a-discussion-of-qualifications-and-skills-in-the-factory-of-the-future-a-german-and-american-perspective

Ghobakhloo, M., Iranmanesh, M., Morales, M. E., Nilashi, M., & Amran, A. (2022). Actions and approaches for enabling Industry 5.0-driven sustainable industrial transformation: A strategy roadmap. *Corporate Social Responsibility and Environmental Management*. https://doi.org/10.1002/csr.2431

Grabowska, S., Saniuk, S., & Gajdzik, B. (2022). Industry 5.0: Improving humanization and sustainability of Industry 4.0. *Scientometrics, 127*(6), 3117–3144. https://doi.org/10.1007/s11192-022-04370-1

Gubernatorov, A., Vinogradov, D., Trofimova, G., & Ulanov, E. (2021). New leadership in digitalization. In E. G. Popkova & B. S. Sergi (Eds.), *Modern global economic system: Evolutional development vs. revolutionary leap* (pp. 1248–1253). Springer.

Hensellek, S. (2019). Digital leadership—Ein rahmenwerk zur erfolgreichen führung im digitalin zeitalter. In T. Kollmann (Ed.), *Handbuch digitale wirtschaft* (pp. 1–19). Springer.

Hensellek, S. (2020). Digital leadership: A framework for successful leadership in the digital age. *Journal of Media Management and Entrepreneurship (JMME), 2*(1), 55–69. https://doi.org/10.4018/JMME.2020010104

Hepaktan, C. E., & Şimşek, D. (2022). Industry 4.0 and the future of the labor market. *İzmir Sosyal Bilimler Dergisi, 4*(2), 80–88. https://doi.org/10.47899/ijss.1174005

Horney, N., Pasmore, B., & O'Shea, T. (2010). Leadership agility: A business imperative for a VUCA world. *Human Resource Planning, 33*(4), 32–38. https://luxorgroup.fr/coaching/wp-content/uploads/Leadership-agility-model.pdf

Huang, S., Wang, B., Li, X., Zheng, P., Mourtzis, D., & Wang, L. (2022). Industry 5.0 and society 5.0—Comparison, complementation and co-evolution. *Journal of Manufacturing Systems, 64*, 424–428. https://doi.org/10.1016/j.jmsy.2022.07.010

Jafari, N., Azarian, M., & Yu, H. (2022). Moving from Industry 4.0 to Industry 5.0: What are the implications for smart logistics? *Logistics, 6*(2), 26. https://doi.org/10.3390/logistics6020026

Jakubik, M., & Berazhny, I. (2017, May 24–27). Rethinking leadership and its practices in the digital era. In *Managing the global economy. Proceedings of the management international conference, Monastier di Treviso, Italy*. University of Primorska Press.

Johansen, B. (2007). *Get there early: Sensing the future to compete in the present.* Berrett-Koehler Publishers.

Karacay, G. (2018). Talent development for industry 4.0. In A. Ustundag & E. Cevikcan (Eds.), *Industry 4.0: Managing the digital transformation* (pp. 123–136). Springer.

Karakose, T., Polat, H., & Papadakis, S. (2021). Examining teachers' perspectives on school principals' digital leadership roles and technology capabilities during the COVID-19 pandemic. *Sustainability, 13*(23), 13448. https://doi.org/10.3390/su132313448

Khalid, K., Chia, S. R., Chew, K. W., & Show, P. L. (2021). Industry 4.0. In P. L. Show, K. W. Chew, & T. C. Ling (Eds.), *The prospect of industry 5.0 in biomanufacturing* (pp. 33–56). CRC Press.

Kin, T. M., & Kareem, O. A. (2019). School leaders' competencies that make a difference in the Era of Education 4.0: A Conceptual Framework. *International Journal of Academic Research in Business and Social Sciences*, *9*(5), 214–225. http://dx.doi.org/10.6007/IJARBSS/v9-i4/5836

Klein, M. (2020). Leadership characteristics in the era of digital transformation. *BMIJ*, *8*(1), 883–902. http://dx.doi.org/10.15295/bmij.v8i1.1441

Kowch, E. (2021). Leading transformation with digital innovations in schools and universities: Beyond adoption. In D. Ifenthaler, S. Hofhues, M. Egloffstein, & C. Helbig (Eds.), *Digital transformation of learning organizations* (pp. 145–168). Springer.

Kulal, A., Divyashree, M. S., & Dinesh, S. (2023). Effectiveness of MOOCs on learning efficiency of students: A perception study. *Journal of Research in Innovative Teaching & Learning*. https://doi.org/10.1108/JRIT-12-2022-0091

Larson, L., & DeChurch, L. A. (2020). Leading teams in the digital age: Four perspectives on technology and what they mean for leading teams. *The Leadership Quarterly*, *31*(1), 101377. https://doi.org/10.1016/j.leaqua.2019.101377

Lee, J., & Kao, A. (2013). Industry 4.0—Factory in big data environment. *Tec. News 26: Strategy*. https://www.researchgate.net/publication/285163697_Industry_40_in_Big_Data_Environment

Leong, Y. K., Tan, J. H., Chew, K. W., & Show, P. L. (2021). Significance of industry 5.0. In P. L. Show, K. W. Chew, & T. C. Ling (Eds.), *The prospect of industry 5.0 in biomanufacturing* (pp. 95–114). CRC Press.

Longo, F., Padovano, A., & Umbrello, S. (2020). Value-oriented and ethical technology engineering in industry 5.0: A human-centric perspective for the design of the factory of the future. *Applied Sciences*, *10*(12), 4182. https://doi.org/10.3390/app10124182

Loy, A. C. M., Chin, B. L. F., & Sankaran, R. (2021). Industrial revolution 1.0 and 2.0. In P. L. Show, K. W. Chew, & T. C. Ling (Eds.), *The prospect of industry 5.0 in biomanufacturing* (pp. 1–14). CRC Press.

Madsen, D. Ø., & Berg, T. (2021). An exploratory bibliometric analysis of the birth and emergence of industry 5.0. *Applied System Innovation*, *4*(4), 87. https://doi.org/10.3390/asi4040087

Maheshwari, S. K., & Yadav, J. (2020). Leadership in the digital age: Emerging paradigms and challenges. *International Journal of Business and Globalisation*, *26*(3), 220–238. https://dx.doi.org/10.1504/IJBG.2020.10032716

Mohebi, L. (2019). Educational leadership and digital culture. *SSRN*. http://dx.doi.org/10.2139/ssrn.3419519

Mokyr, J. (2000). Innovation and its enemies: The economic and political roots of technological inertia. In M. Olson, & S. Kähköhnen (Eds.), *A not-so-dismal science: A broader view of economies and societies* (pp. 61–91). Oxford University Press.

Mosconi, F. (2015). *The new European industrial policy: Global competitiveness and the manufacturing renaissance*. Routledge.

Nahavandi, S. (2019). Industry 5.0—A human-centric solution. *Sustainability, 11*(16), 4371. https://doi.org/10.3390/su11164371

Oberer, B., & Erkollar, A. (2018). Leadership 4.0: Digital leaders in the age of industry 4.0. *International Journal of Organizational Leadership*. https://ssrn.com/abstract=3337644

Østergaard, E. H. (2018). *Welcome to industry 5.0*. https://info.universal-robots.com/hubfs/Enablers/White%20papers/Welcome%20to%20Industry%205.0_Esben%20%C3%98stergaard.pdf

Paschek, D., Mocan, A., & Draghici, A. (2019). Industry 5.0—The expected impact of next industrial revolution. In *Thriving on future education, industry, business, and society, proceedings of the make learn and TIIM international conference*, Piran, Slovenia.

Raj, A., & Jeyaraj, A. (2023). Antecedents and consequents of Industry 4.0 adoption using technology, organization and environment (TOE) framework: A meta-analysis. *Annals of Operations Research*, *322*(1), 101–124. https://doi.org/10.1007/s10479-022-04942-7

Raja Santhi, A., & Muthuswamy, P. (2023). Industry 5.0 or Industry 4.0 S? Introduction to Industry 4.0 and a peek into the prospective Industry 5.0 technologies. *International Journal on Interactive Design and Manufacturing (IJIDeM)*, 1–33. https://doi.org/10.1007/s12008-023-01217-8

Rifkin, J. (2011). *The third industrial revolution: How lateral power is transforming energy, the economy, and the world*. Palgrave Macmillan.

Sachsenmeier, P. (2016). Industry 5.0—The relevance and implications of bionics and synthetic biology. *Engineering, 2*(2), 225–229. http://dx.doi.org/10.1016/J.ENG.2016.02.015

Salkin, C., Oner, M., Ustundag, A., & Cevikcan, E. (2018). A conceptual framework for Industry 4.0. In A. Ustundag, & E. Cevikcan (Eds.), *Industry 4.0: Managing the digital transformation* (pp. 3–23). Springer.

Schneider, P. (2018). Managerial challenges of Industry 4.0: An empirically backed research agenda for a nascent field. *Review of Managerial Science, 12*(3), 803–848. https://doi.org/10.1007/s11846-018-0283-2

Sigov, A., Ratkin, L., Ivanov, L. A., & Xu, L. D. (2022). Emerging enabling technologies for industry 4.0 and beyond. *Information Systems Frontiers*, 1–11. https://doi.org/10.1007/s10796-021-10213-w

Sheninger, E. (2019). *Digital leadership: Changing paradigms for changing times*. Corwin Press.

Sony, M. (2020). Pros and cons of implementing Industry 4.0 for the organizations: A review and synthesis of evidence. *Production & Manufacturing Research*, *8*(1), 244–272. https://doi.org/10.1080/21693277.2020.1781705

Sung, T. K. (2018). Industry 4.0: A Korea perspective. *Technological Forecasting and Social Change*, *132*, 40–45. https://doi.org/10.1016/j.techfore.2017.11.005

Tan, C. H. (2021). Industry 3.0. In P. L. Show, K. W. Chew, & T. C. Ling (Eds.), *The prospect of Industry 5.0 in biomanufacturing* (pp. 33–56). CRC Press.

Tigre, F. B., Curado, C., & Henriques, P. L. (2023). Digital leadership: A bibliometric analysis. *Journal of Leadership & Organizational Studies*, *30*(1), 40–70. https://doi.org/10.1177/15480518221123132

Tiwari, S., Bahuguna, P. C., & Walker, J. (2022). Industry 5.0: A macroperspective approach. In V. Garg, & R. Goel (Eds.), *Handbook of research on innovative management using AI in industry 5.0* (pp. 59–73). IGI Global.

World Economic Forum. (2020). *The future of jobs report 2020*. https://www3.weforum.org/docs/WEF_Future_of_Jobs_2020.pdf

Yang, F., & Gu, S. (2021). Industry 4.0, a revolution that requires technology and national strategies. *Complex & Intelligent Systems*, *7*, 1311–1325. https://doi.org/10.1007/s40747-020-00267-9

Dr. Sezin Esfer

Lifelong Learner Leaders

Abstract

This chapter stands for the necessity of a lifelong learning approach in the context of the leaders' adaptability and sustainability for normative reflection, both about the norms in the VUCA world and their own role. We need this reflection because in a world where change, transformation, and complexity are the leading actors, how we perceive life, and our learning processes are continuously renewed (Bozkurt, 2019). This dynamic nature of the world is defined as VUCA, which can be a great opportunity for leaders' professional development by helping them with the needed skills, the current situation, and what should be done by showing the challenges and threats of the unpredictability (Raja, 2021). On the other hand, just like scientists, leaders also need to prepare for the future with great care, make their goals, values, and visions explicit, and transparently consider the co-production of knowledge (Büttner et al., 2023). By being lifelong learners, leaders can develop the agility, resilience, and creativity needed to navigate the uncertainty and complexity of the VUCA world. Similarly, in Industry 5.0, where the lines between humans and machines blur, and technology is integrated with work processes, leaders need to constantly learn and adapt to leverage the potential of technology and human intelligence. That's why this chapter provides up-to-date information on the challenges that educational leaders face in developing the digital skills needed to lead in a VUCA world and emphasizes the importance of ongoing professional development to address this issue. Specifically, the three main themes: (1) The skill gap phenomenon, (2) Leadership in Open Distance Digital Education context, and (3) Massive Open Online Courses as lifelong learner supporters highlight the lifelong learning solutions within the educational and business world context for effective, agile, and digital leadership. Last, some tips, suggestions, and research findings for today and future leaders are shared.

Keywords: digital leadership, VUCA, lifelong learning, Massive open online courses (MOOC), open distance digital education (ODDE)

Leaders who can embrace the VUCA world by learning new skills, staying abreast of emerging technologies, and cultivating a culture of innovation can unlock new levels of productivity, efficiency, and satisfaction. They can offer radical changes within the whole system for sustainable development (Büttner et al., 2023). So, this chapter specifically provides key concepts of being a digital and lifelong learner leader in a VUCA world and practical suggestions for leaders to navigate through uncertainty and change. So, this chapter aims to focus on three main issues to provide a holistic approach to learning needs and how they can be met. Firstly, the skill gap phenomenon between universities and the business world, learning needs and prerequisite knowledge, and digital and social skills are examined. Secondly, *Open, Distance, and Digital Education (ODDE) mainstream as a solution for professional development* is discussed and lastly, Massive Open Online Courses (MOOCs) as lifelong learning supporters are clarified. In connection with these three issues, basically, it can be said that the aim is to introduce readers to what kind of skills are needed for effective leadership, how to overcome skill gap problems, how to manage a digitalized world, and lastly how MOOCs can help us in this framework.

The Skill Gap Phenomenon

Firstly, we should admit that no one today has the luxury of staying away from the needs and capabilities of the twenty-first century, digitized systems, and Industry 5.0 (Ciolacu et al., 2023). Whereby the "earning = learning" approach gained importance in a complex and digitalized world and the formula "a highly educated workforce + basic research + commercial support = innovative products + new companies + jobs + prosperity" emerged (Coates, 2013, p. 3) so long ago. According to the Strategic Intelligence Report of Word Economy Forum (WEF), while it is estimated that over 80 million jobs will be lost by 2025, over 100 million jobs may have emerged (WEF, 2022). The rapid pace of technological change is creating a skills gap in many industries, including education. According to the WEF (2022) Report, today, most of the children starting primary school may have jobs that we do not know yet. This estimation highlights that educators and educational leaders need to improve their digital skills in order to prepare for the future. Since the cooperation of humans, machines, and algorithms causes this situation, WEF (2020) also highlights the importance of the skills of creativity, curiosity, leadership, flexibility, and technological competencies. As agility, resilience, and sustainability are the focus skills in the VUCA world (Raja, 2021), as a global issue, new-generation leaders are the key people who can take responsibility and create policies regarding technology access and management (WEF, 2023).

As the spiral structure of leadership shows crucial concepts (adaptive leadership, technology leadership, etc.) related to leadership, it may also herald the increasing gap between the skills needed and the skills possessed (see Fig. 1). The gap between soft skills for employees and employers already becomes clear (DeakinCo, 2017) a long time ago. On the other hand, the considerable need for transferable skills (problem-solving, entrepreneurship, and creative thinking) is still enhancing (Becker et al., 2017; Santandreu Calonge et al., 2019). Moreover, the International Society for Technology in Education (ISTE) identified digital leadership as a key competency for educational leaders and emphasized the need for ongoing training and professional development to develop these skills (ISTE, 2022). When we specifically examine the ISTE digital leadership standards, which are "Excellence in Professional Practice, Visionary Leadership, Digital-age Learning Culture, Digital Citizenship, and Systemic Improvement," they provide continuous improvement in their school environment by effectively managing resources and technology.

Learning Needs

Most educational leaders lack the necessary digital skills to effectively implement digital transformation initiatives, and that ongoing professional development is essential to address this issue. According to National School Boards Association, most of the school board members do not believe that their educational leaders have the necessary technical skills to effectively lead digital transformation (NSBA, 2022). Similarly, according to the Consortium for School Networking (CoSN), that is the leading professional association for school technology leaders, the Driving K-12 Innovation report (2023) state that leaders cannot create effective digital ecosystems in schools where the information technology resource enable students and teachers to communicate effectively. This indicates a significant gap between the skills that educational leaders believe they have and the skills that are required to effectively manage digital transformation in a rapidly changing environment. Furthermore, global challenges related to twenty-first Century requirements (Sablina et al., 2018; Zawacki-Richter et al., 2018) and the increasing value of informal learning for digitalization in the workplace (Egloffstein, 2018; Santandreu Calonge et al., 2019) also create the need for ongoing professional development and training to help educational leaders acquire the necessary digital skills and knowledge to effectively manage digital transformation in a VUCA world. That's why leadership is affected by various learning needs and dimensions in Fig. 1.

The Need for Informal Learning

Today, there is a large gap between expected job skills and formal education skills (Patru & Balaji, 2016) and varying learning needs that must be met quickly (De Vries & Brall, 2008). In other words, as the gap between university degrees, graduates, education systems, and industry needs grows (Patru & Balaji, 2016), there must be continuous personal development for all people working to adapt

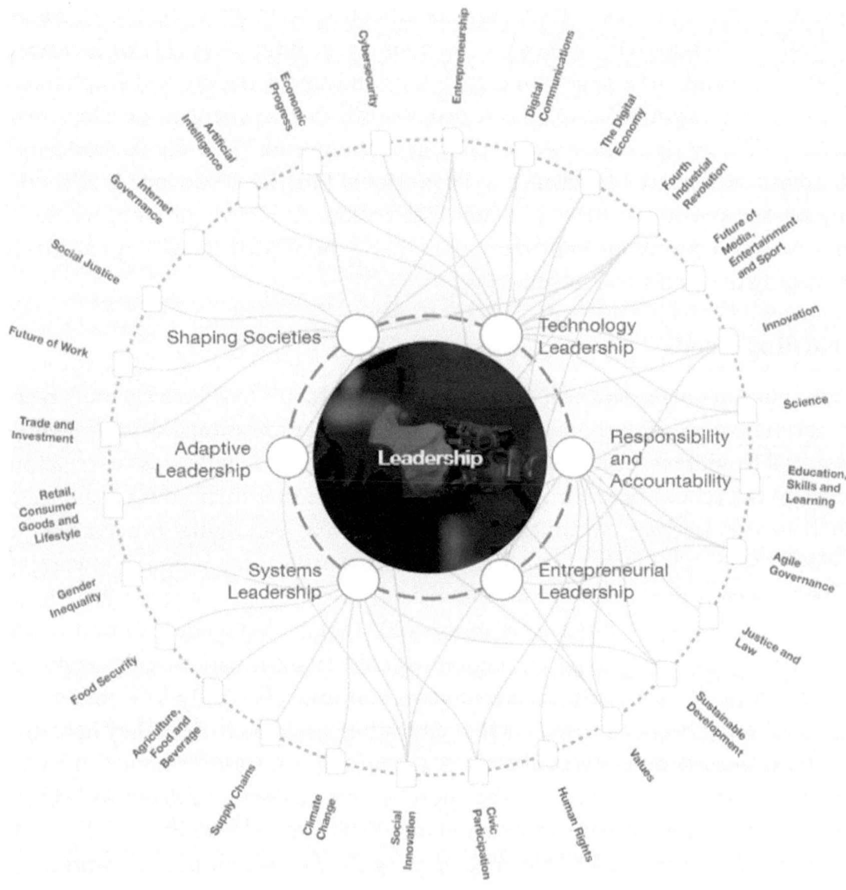

Fig. 1. Leadership

Source: https://intelligence.weforum.org/topics/a1G0X000004Q9aRUAS

to such new job demands and technologies (Small et al., 2019). Similarly, the value of affirming non-formal and informal learning environments to promote employability, mobility and support socioeconomically disadvantaged people is underlined:

> Employers are looking for "work-ready" graduates who can apply their academic studies and abilities in a commercial or work context. The industry is changing at a rapid rate. This presents a dilemma for universities and colleges if they try to keep up with industry

> demands. Graduates need to upskill and adapt to a changing jobs market. Their degree will only get them in so far in a career that may span 50 years. (Santandreu Calonge et al., 2019, p. 19)

Unfortunately, traditional education is not sufficient and efficient to meet the mentioned demands of "twenty-first-century competencies" (Egloffstein, 2018). On the other hand, working adults and leaders, in general, cannot take classes during working hours (Condé & Cisel, 2019). Even many people do not believe they can improve their capacity and job skills (Hirschi, 2018; Small et al., 2019). De Vries and Brall (2008) argue that providing institutions with informal learning is a significant opportunity as a natural and flexible form of learning can support more employees or leaders. Informal learning can be regarded as the result of daily work-related activities and is not structured by disciplines (Witthaus et al., 2016).

Leadership Skills in the VUCA World

When the reasons for adults' learning demands are examined (see Fig. 2), three continuous streams of *workplace learning, informal learning, and lifelong learning* become visible. Because adults and professionals as lifelong learners need to be adapted to their constantly changing learning needs (Castaño-Muñoz et al., 2017; De Vries & Brall, 2008; Eşfer & Cagiltay, 2018). Specifically, the leadership profession requires a set of skills and qualities to survive and sustain the impact of the VUCA factors (Sinha & Sinha, 2020). In other words, leadership in a VUCA world requires constant learning, adaptability, collaboration, innovation, and experimentation so they are required to navigate uncertainties, paradoxes, conflicts, pressures, and ambiguities. Likewise, critical factors for leaders to be successful in the VUCA world include responsibility, agility, networking, innovation, and ethical practices (Sarkar, 2016).

All issues mentioned before like "the skills gap phenomenon," "the need for informal learning," "job-related skills," and "informal learning" point out that Open, Distance, and Digital Education (ODDE) as a solution for professional development. Also, being a digital leader positively affects the teachers' technology integration in the school environment (AlAjmi, 2022; Damayanti & Mirfani, 2020). Moreover, the Council of the European Union (2019) adopts recommendations on key competencies for lifelong learning such as competence in science, technology, and engineering, digital competence, learning-to-learn competence, citizenship competence, and entrepreneurship competence. Overall, the statistics and the related literature suggest that there is a pressing need for educational

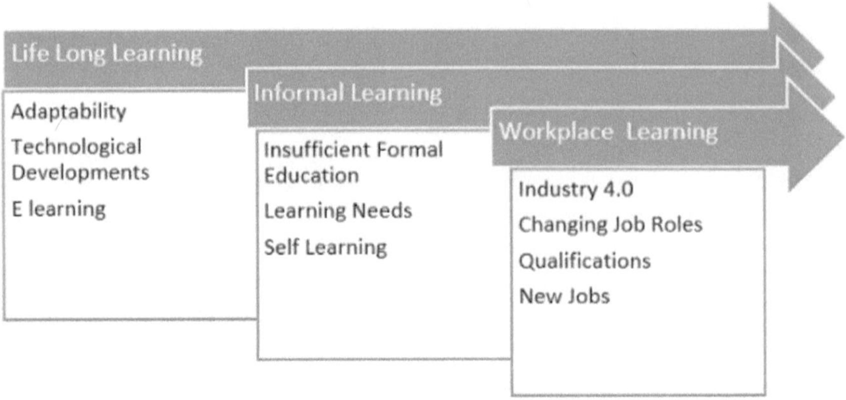

Fig. 2. Reasons for Learning Demands (Eşfer & Cagiltay, 2018)

leaders to improve their digital skills and knowledge to effectively navigate the challenges of the VUCA world.

Developing Digital Leadership in the ODDE Context

ODDE could be the "mainstream" of learning and teaching (Xiao, 2018), then it became the normal of education due to the pandemic (Bond et al., 2021) and finally today ODDE has an undeniable impact on our lives along with the e bias against it (Xiao, 2022). The pandemic created an urgent and extra need for distance and digital learning (AlAjmi, 2022). Although online and active learning can increase student satisfaction (Nasir, 2020), educators and educational leaders faced new difficulties that no researchers could predict during this period (Zimmer et al., 2021). On the other hand, it is possible to claim that lifelong learning is essential for leadership development in a VUCA world. If leaders are committed to continuous learning and development. MOOCs can provide a flexible and accessible way for leaders to engage in lifelong learning and professional development. They offer a wide range of courses and learning opportunities in various fields, which can help leaders stay up to date with emerging trends and new skills. Leaders need to be adaptable and responsive to change in a VUCA world. This requires a commitment to continuous learning and development throughout their careers. Likewise, digital literacy is an essential skill for educational leaders in open learning environments.

Digital Leadership

Digital leadership is a broader concept that encompasses the strategic and visionary use of technology while using technology effectively in leadership is more tactical and operational in nature. Digital leadership is very crucial in terms of providing the role of educational leaders with a strategic perspective on how educational needs can be met (AlAjmi, 2022; Hickman & Akdere, 2018). Both are important for leaders to master in today's digital world. Because the digital transformation of education has become a necessity in the twenty-first century and the widespread use of technology in education has created an open learning environment that requires educational leaders to embrace new approaches to teaching and learning. Educational leaders with a high level of digital leadership use technology more effectively in both administrative and instructional work and thus can affect the communication of every stakeholder in the school environment (AlAjmi, 2022). Educational leaders are expected to have the necessary competencies in digital literacy (Christopoulous et al., 2021). On the other hand, unfortunately, many leaders are not willing to improve their digital literacy skills (AlAjmi, 2022). The International Society for Technology in Education (ISTE) suggests five aspects of digital leadership (ISTE, 2014): "Excellence in Professional Practice, Visionary Leadership, Digital age Learning Culture, Digital Citizenship, and Systemic Improvement."

a. *Excellence in Professional Practice:* An educational leader should aim to enable students to learn through their digital and technological resources. They can also foster innovation and learning environments for professional development.
b. *Visionary Leadership:* The main purpose of evaluating visionary leadership within the framework of digital leadership is to increase the effectiveness of decision-making processes. On the other hand, an education leader should have up-to-date knowledge of technology trends to articulate his vision for improvement when communicating with school stakeholders.
c. *Digital-Age Learning Culture:* An educational leader must create the resources and connections necessary in the age of digital learning and promote excellence. Digital leaders should also raise awareness of the strategies that form the framework for implementing a digital-age learning culture in schools.
d. *Digital Citizenship:* Fulfilling ethical, legal, and social duties as an education leader develops digital culture and citizenship. In particular, the spread of digital citizenship to digital leadership is important in this context.

e. *Systemic Improvement:* Basically, an educational leader must ensure continuous improvement in her school and effectively manage the resources, technical knowledge, and skills required for this.

MOOCs for Digital Leadership in the VUCA World

The world of VUCA can provide opportunities for leaders to develop a clear vision, develop a deep understanding of their environment, communicate clearly, and maintain agility in the face of rapid change. Being a lifelong learner is critical for leaders to succeed in a VUCA world where change is constant and innovation is essential and Industry 5.0. Continuously learning leaders can stay ahead of the curve, adapt to changing circumstances, and drive growth and success for their organizations. Leaders must be lifelong learners in the VUCA world and Industry 5.0 because these environments require continuous adaptation, innovation, and learning to be successful. In a VUCA world, the pace of change is accelerating and new technologies, business models, and societal trends are constantly emerging. Leaders who fail to keep up with these changes risk being apathetic and missing out on opportunities for growth and innovation. Digital technologies especially mean the availability of resources, flexibility, learning environments at their own pace, and controlling learning for leaders (Santandreu Calonge et al., 2019). At this point, MOOCs took the stage, promising to quench the thirst for high-quality, personalized, and free education worldwide (Kay et al., 2013).

MOOCs can assist students in the immediate learning needs of their profession in the "rapidly changing workforce" of the twenty-first century (Santandreu Calonge et al., 2019), and the use of digital technologies in workplace learning enables informal and lifelong learning (Egloffstein & Ifenthaler, 2017) at the same time. While MOOCs support more personalized learning experiences (Egloffstein, 2018), unlike professional education, they often offer an informal learning environment and contribute to professional development (Condé & Cisel, 2019; Haemer & Borges-Andrade, 2018; Moraes & Borges-Andrade, 2015). The opportunity to use digital technologies in workplace learning provides informal and lifelong learning (Egloffstein & Ifenthaler, 2017). Due to their flexible nature, MOOCs offer universal and inclusive learning environments for everyone (Çelik et al., 2023) and are important for digital workplace learning (Egloffstein & Ifenthaler, 2017; Egloffstein et al., 2019). Moreover, the lifelong learning platform declared, of MOOCs combats inequalities in education and serves the comprehensive social justice struggle (Patru & Balaji, 2016).

MOOCs as Lifelong Learning Supporters

MOOC is still a much-talked-about phenomenon in the 2020s (Bonvillian & Singer, 2013; Liyanagunawardena et al., 2013) and can be seen as a new teaching/ learning approach (Guardia et al., 2013). MOOCs are free for everyone around the world (Cormier & Siemens, 2010; Kop & Carroll, 2011) with videos, quizzes, resources, articles, discussions, etc. for an unlimited learning community. They can be defined as a combination and integration (Ho et al., 2015). "The letter 'M' in the MOOC abbreviation denotes being large and diverse in cultural, socioeconomic, demographic and many other dimensions" (Bozkurt et al., 2017, p. 138). "Openness can be considered the biggest challenge for MOOCs and their quality. It means more than free and open access: it can be about open approaches, that is, using innovative approaches to learning and education" (Walia, 2020, p. 18). Unfortunately, we cannot claim that many MOOCs are "open" as they should be (Eşfer, 2020). On the other hand, the lack of entry requirements such as age, location, or previous formal education and free accessibility are among the most important advantages of MOOCs (Patru & Balaji, 2016).

When we examine the leaders' perspectives on lifelong learning and MOOCs, it is purely clear how leaders in the technology industry and all other industries recognize the importance of lifelong learning and that's why they are using MOOCs as a tool for personal and professional development. For the first example, Jeff Weiner, CEO of LinkedIn, wrote, "I've long believed that investing in yourself is the best investment you can make. MOOCs have made that investment increasingly accessible to everyone." in a LinkedIn post. Second, Sundar Pichai, CEO of Google, encouraged employees to take MOOCs to continue learning and developing their skills. Third, Satya Nadella, the CEO of Microsoft, has spoken publicly about the importance of continuous learning and has encouraged his employees to take advantage of MOOCs and other learning opportunities. Microsoft also offers its own online learning platform, Microsoft Learn, which includes a variety of courses and certifications. Fourthly, Jack Welch, the former CEO of General Electric, was known for his commitment to lifelong learning by encouraging his employees to take risks and learn from their failures.

MOOCs can be seen as a potential solution for leadership development in a VUCA world, by offering flexible and accessible learning opportunities that can help leaders develop the key competencies and strategies needed to navigate rapidly changing and uncertain environments. On the other hand, MOOCs for professional development (pdMOOCs) were designed and developed through an EU Project which is called bilgeis.net (Eşfer & Cagiltay, 2018). Although bilgeis.

net, one of the largest Turkish MOOC platforms with more than 750 thousand users, stands for technical skills, the most popular courses are Stress Management, Leadership, Coping with Difficult People, etc. in the platform (Bozkurt et al., 2021; Cagiltay et al., 2020). In addition, MOOCs on social skills such as "Using Body Language Correctly, Mediation, and Effective Time Management" are frequently preferred in Akadema, which is also the first and favorite MOOC platform in Turkey (Bozkurt et al., 2021). From an international perspective, according to ChatGPT results and Class Central, the following MOOC portals offer a wide range of courses and programs from various universities and organizations, covering diverse subjects such as computer science, business, arts, and humanities, among others (ChatGPT, 2023): Coursera (www.coursera.org), edX (www.edx.org), Udacity (www.udacity.com), FutureLearn (www.futurelearn.com), LinkedIn Learning (www.linkedin.com/learning), Skillshare (www.skillshare.com), Codecademy (www.codecademy.com), Khan Academy (www.khanacademy.org), OpenLearning (www.openlearning.com), and Alison (alison.com). You can find a short list of international MOOC portals that all leaders can get benefit from in terms of professional development and specifically being an effective leader in the VUCA world in Tab. 1.

Tab. 1. International MOOC Portals (Class Central, 2023)

MOOC Platforms from North America & United States	Coursera by the United States edX by the United States Udacity by the United States Canvas Network by United States Kadenze by the United States Stanford Lagunita by the United States Complexity Explorer by United States MéxicoX by Mexico
MOOC Platforms from Europe & United Kingdom	FutureLearn by the United Kingdom France Université Numérique (FUN) by France Miríadax by Spain EduOpen by Italy Federica Web Learning by Italy European Multiple MOOC Aggregator (EMMA)—Europe OpenHPI by Germany MOOC.fi by Finland Prometheus by Ukraine Open Education (openedu.ru) by Russia

MOOC Platforms from Asia (excluding Chinese MOOC Platforms)	SWAYAM by India NPTEL by India JMOOC by Japan gacco by Japan OpenLearning by Japan K-MOOC by Korea ThaiMOOC by Thailand Edraak (Arabic) by Jordan Campus-Il by Israel
Chinese Language MOOC Platforms	XuetangX by China Chinese University MOOC by China Zhihuishu by China CNMOOC by China Xue Yin Online by China Open Education (openedu.tw) by Taiwan eWant by Taiwan

It is obvious that there are many more MOOC providers (institutions and organizations) apart from these portals placed in the table above. For instance, Harvard University offers a range of free online courses through its HarvardX platform, allowing learners to access its world-class curriculum from anywhere in the world. If we examine the necessary conditions to get the best efficiency from the MOOC portals mentioned, some key concepts become visible for lifelong learning leaders such as self-directed learning, collaborative learning, microlearning, learning agility, adaptation and change, and innovation Fig. 3 summarizes some of the related concepts with using MOOCs as lifelong supporters of leaders to survive in the VUCA world.

Self-directed Learning: MOOCs can empower leaders to take ownership of their learning and development. MOOCs offer a self-paced learning environment, which allows leaders to learn at their own pace and focus on the topics and skills that are most relevant to their professional goals. Self-directed learning can be a powerful tool for leadership development in a VUCA world. MOOCs offer suggestions for how to incorporate self-directed learning into leadership development programs because leaders who take ownership of their own learning and development are more likely to be proactive, adaptable, and innovative.

Collaborative Learning: In MOOC portals, instructors, faculty members, and participants have the opportunity to interact with a global network of peers through the course's online platform. During the 2008 financial crisis, leaders like Jamie Dimon (JP Morgan Chase) and Lloyd Blankfein (Goldman Sachs) were able to guide their companies through a turbulent period by making tough

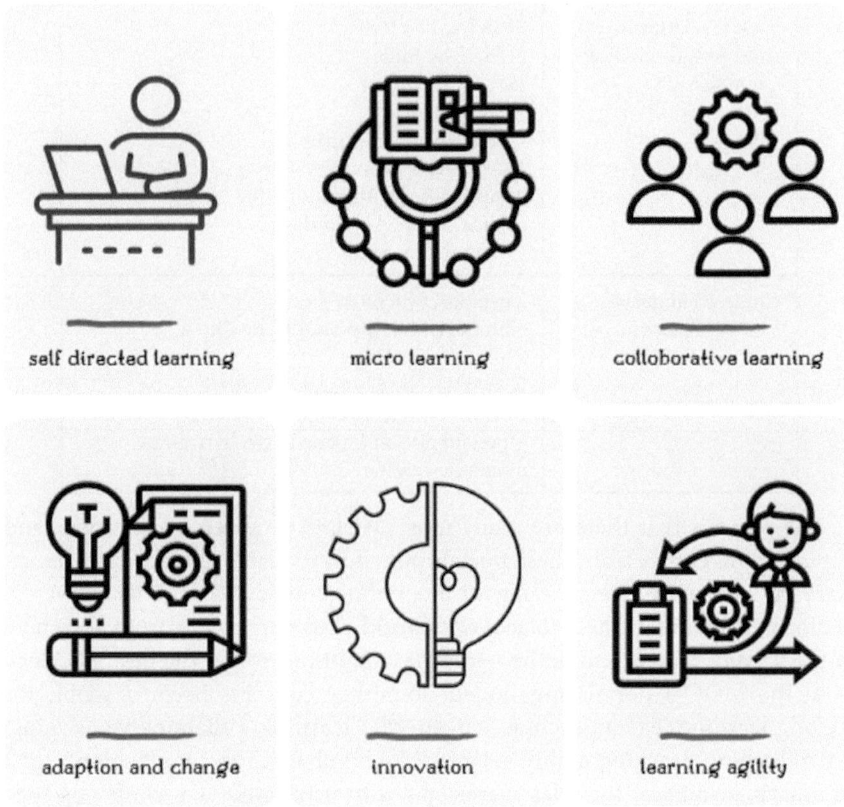

Fig. 3. Some Related Concepts with Using MOOCs as Lifelong Supporters of Leaders to Survive in the VUCA World

decisions and communicating a clear vision for the future. So, educational leaders also should create opportunities for collaboration and cooperation among teachers, administrators, and other stakeholders. MOOCs can provide leaders with opportunities to connect with peers and colleagues from around the world. MOOCs often feature interactive discussion forums and collaborative learning activities, which can help leaders build strong relationships with other professionals in their fields. Spotify encourages collaboration between its teams and focuses on delivering new features and updates to users quickly and efficiently.

Microlearning: Microlearning basically refers to a learning approach to delivering content in small, bite-sized chunks. Microlearning is often delivered online and can be accessed on demand. IBM has used microlearning to support lifelong learning among its employees. The company has developed a mobile learning app called IBM Think Academy, which offers short videos and articles on a variety of topics related to digital leadership. Similarly, The Division of Diversity and Community Engagement at the University of Texas at Austin has implemented a microlearning program to support their educational leadership initiatives. The program, called "Tools for Diversity and Inclusion," offers short, on-demand videos that cover topics such as unconscious bias, cultural competency, and inclusive teaching practices.

Learning Agility: Learning agility is a key competency for leadership in a VUCA world. Learning agility has become a key component of leadership success, and it is essential to take action in new situations in today's VUCA world. In a VUCA world, leaders need to be able to navigate ambiguity and make decisions quickly. Learning agility allows them to adapt to changing circumstances and learn from their experiences. So, there are several MOOCs available online that cover the topic of leadership in the VUCA world for leaders at all levels, from CEOs and senior executives to mid-level managers and emerging leaders. Some of them examine the impact of digital disruption on leadership, providing learning agility, and offering strategies for navigating this new environment while some of them focus on how to lead in a VUCA world.

Innovation: MOOCs can provide leaders with exposure to new ideas and approaches to problem-solving. MOOCs often feature courses and learning opportunities in emerging fields and technologies, which can help leaders stay abreast of the latest developments in their industries and foster a culture of innovation. Leaders who foster a culture of innovation, encourage experimentation, and take calculated risks are more likely to thrive in rapidly changing environments. Many well-known leaders from education and different sectors encouraged employees to take MOOCs and other online courses to follow the trends and issues in the professional areas. By embracing disruptive technologies and leveraging data to inform decision-making, educational leaders can drive innovation in education and create new opportunities for learners to succeed.

Adaptation and Change: During the pandemic, many businesses were forced to adapt quickly to new challenges. Leaders who were able to pivot their strategies and adapt to changing circumstances were more likely to succeed. For example, many restaurants started offering takeout and delivery services to make up for lost dine-in revenue. Mary Barra, CEO of General Motors, has spoken about the need for continuous learning and has encouraged employees to take advantage of educational opportunities by saying "The pace of change in the auto industry is accelerating, and we need to be continuously learning and adapting to stay ahead." in a speech to the Society of Automotive Engineers. For example, while the pandemic has accelerated the adoption of online learning, educational leaders have had to adapt quickly to this new reality. For example, the University of South Africa has been offering distance education for over 70 years but had to rapidly transition to online learning during the pandemic. This required leaders to develop new skills and strategies to ensure that students could continue their studies remotely.

Conclusion

This chapter aims to reflect a review of the literature on how to become lifelong learner leaders and the importance of lifelong learning in developing leadership skills in a VUCA world. So, we can identify key competencies and strategies for developing leadership skills in a VUCA environment and discuss the challenges and opportunities associated with leadership development in a rapidly changing world. Based on the findings and outcomes of the studies provided, MOOCs can be seen as a potential solution for developing the key competencies and strategies needed for leadership development in a VUCA world. Because whereby the "earning = learning" approach gained importance in a complex and digitalized world and the formula "a highly educated workforce + basic research + commercial support = innovative products + new companies + jobs + prosperity" emerged (Coates, 2013, p. 3) so long ago, then there is a considerable need for transferrable skills (social, intercultural, and digital skills) in the rapidly changing world (Becker et al., 2017; Santandreu Calonge et al., 2019). However, to adapt to this kind of new job demands and skills, there should be continuous self-improvement in human capacity (Small et al., 2019). Since any kind of *production* task cannot be stopped in a workplace such as factories, firms, institutions, schools, etc. and the leaders have no time for on-the-job training and very limited time/opportunity to learn new skills. In that point, MOOCs can help learners in terms of learning needs which should be immediately met for their job in the twenty-first century's "fast-changing workforce" (Santandreu Calonge et al., 2019) and using digital technologies in workplace learning provides informal and lifelong learning (Egloffstein & Ifenthaler, 2017) at the same time. The following suggestions offer additional insights and perspectives on leadership in the VUCA world. By referencing the MOOCs, the chapter can provide a robust and evidence-based discussion of the topic:

- Educational leaders should also encourage the integration of digital technology into the curriculum since they should be better equipped to navigate rapidly changing environments and respond to emerging challenges. Radical decisions and changes can be taken into account within the whole system beginning from the production to the assessment of the knowledge if we want to produce the knowledge needed for sustainable development (Büttner et al., 2023).

- Being a digital leader can positively affect the teachers' technology integration in the school environment, and leaders can question paradigms regarding the nature of learning and learning needs in terms of their professional development. Also, they should embrace change and be willing to adapt to new technology and pedagogy. They should be open to feedback and use it to improve the digital transformation process.
- Educational leaders should develop a strategic vision that outlines the goals and objectives of digital transformation in open learning environments. This vision should be aligned with the mission and values of the educational institution. Managing digital transformation in open learning environments requires educational leaders to develop a strategic vision, build a culture of innovation, foster collaboration, promote digital literacy, and embrace change.
- ISTE's suggestions (2014) based on the five aspects of digital leadership (Excellence in Professional Practice, Visionary Leadership, Digital-Age Learning Culture, Digital Citizenship, and Systemic Improvement) should be considered.
- MOOCs show how lifelong learning can help leaders stay adaptable and responsive to changing environments and offer suggestions for how to incorporate lifelong learning into leadership development programs. Because they have the potential to provide learners with access to high-quality education and training, regardless of their location or financial resources. MOOC portals suggestions provided in the chapter are essential for educational leaders to manage digital transformation and improve themselves.
- It can be searched "Digital Leadership," "Leading in the Digital Age," and "Digital Transformation and Leadership" keywords in these MOOC portals. There are some MOOCs related to digital leadership, VUCA world, which cover the challenges of leadership in complex environments and provide strategies for digital disruption, leading change, and making decisions, and offer strategies for navigating this new environment. Leading Through Digital Disruption offered by Coursera, Leading in a Complex Environment offered by edX.
- In addition to formal training and development programs, leaders should take ownership of their own learning and development by seeking out new opportunities for growth and development.

- Leaders should focus on building strong relationships with their team members, stakeholders, and other key players in their industry to navigate uncertain and rapidly changing environments. Also, the community of practice can be a significant way for new skills.
- Leaders should seek out new experiences, be open to feedback, and learn from their mistakes to build their learning agility. Leaders who can learn quickly and adapt to new situations are more likely to be successful in uncertain and rapidly changing environments. MOOCs can provide leaders with opportunities to learn quickly and adapt to new situations. MOOCs often feature interactive and experiential learning activities, which can help leaders build their learning agility by practicing new skills in a safe and low-risk environment.
- Learning agility, the culture of innovation, self-directed learning, microlearning, and collaborative learning are some of the key concepts in terms of needed competencies for leadership development in a VUCA world. Also, they are essential for successful digital transformation in open learning.

References

Antonacopoulou, E. (2018). Organisational learning for and with VUCA: Learning leadership revisited. *Teoria e Práctice em Administração [Theory and Practice Management Journal–Leading Brazilian Journal]* (Special Issue), 8(2), 10–32.

AlAjmi, M. K. (2022). The impact of digital leadership on teachers' technology integration during the COVID-19 pandemic in Kuwait. *International Journal of Educational Research, 112*, 101928.

Bansal, A., Panchal, T., Jabeen, F., Mangla, S. K., & Singh, G. (2023). A study of human resource digital transformation (HRDT): A phenomenon of innovation capability led by digital and individual factors. *Journal of Business Research, 157*, 113611.

Becker, S. A., Cummins, M., Davis, A., Freeman, A., Hall, C. G., & Ananthanarayanan, V. (2017). *NMC horizon report: 2017 higher education edition* (pp. 1–60). The New Media Consortium.

Bozkurt, A. (2019). Vizyon 2023: Türkiye'de açık ve uzaktan öğrenme alanında somut ve soyut teknolojiler bağlamında eğilimler. *AUAd, 5*(4), 43–64.

Bozkurt, A., Kocdar, S., Cagiltay, K., Esfer, S., Celik, B., Karaman, S., & Kursun, E. (2021). Massive Open Online Courses (MOOCs) in Turkey and an analysis in the context of turkish higher education. *Journal of Higher Education, 11*(2 Pt 2), 521–536.

Bonvillian, W. B., & Singer, S. R. (2013). The online challenge to higher education. *ISSUES in Science and Technology*, *29*(4), 23–30.

Büttner, L., Darbi, M., Haase, A., Jax, K., Lepenies, R., Priess, J., & Zeug, W. (2023). Science under pressure: How research is being challenged by the 2030 Agenda. *Sustainability Science*, 1–6. https://doi.org/10.1007/s11625-023-01293-5

Cagiltay, K., Esfer, S., & Celik, B. (2020). Insights into a nationwide pdMOOC portal: Bilgeis.net of Turkey. In K. Zhang, C. J. Bonk, T. C. Reeves & T. H. Reynolds (Eds.), *MOOCs and open education in the global south: Challenges, successes, and opportunities* (pp. 130–139). Routledge. https://doi.org/10.4324/9780429398919

Castaño-Muñoz, J., Kreijns, K., Kalz, M., & Punie, Y. (2017). Does digital competence and occupational setting influence MOOC participation? Evidence from a cross-course survey. *Journal of Computing Higher Education*, 29, 28–46. https://doi.org/10.1007/s12528-016-9123-z

Celik, B., Esfer, S., & Cagiltay K. (2023). Expert heuristic collaborative approach for usability testing in a MOOC portal. *New Design Ideas*, *7*(1), 81–94.

Christopoulos, A., & Sprangers, P. (2021, January 1). Integration of educational technology during the Covid-19 pandemic: An analysis of teacher and student receptions. *Cogent Education*, *8*(1), 1964690. https://doi.org/10.1080/2331186x.2021.1964690

Ciolacu, M. I., Mihailescu, B., Rachbauer, T., Hansen, C., Amza, C. G., & Svasta, P. (2023). Fostering engineering education 4.0 paradigm facing the pandemic and VUCA world. *Procedia Computer Science*, *217*, 177–186.

Class Central. (n.d.). *MOOC list*. Retrieved April 15, 2023, from https://www.classcentral.com/report/mooc-platforms/

Coates, K. (2013). The re-invention of the academy: How technologically mediated learning will—and will not—transform advanced education. In S. K. S. Cheung, J. Fong, W. Fong, F. L. Wang, & L. F. Kwok (Eds.), *Hybrid learning and continuing education. ICHL 2013* (Lecture Notes in Computer Science, Vol. 8038). Springer. https://doi.org/10.1007/978-3-642-39750-9_1

Condé J., & Cisel, M. (2019) On the use of MOOCs in companies: A panorama of current practices. In M. Calise, C. Delgado Kloos, J. Reich, J. Ruiperez-Valiente & M. Wirsing (Eds.), *Digital education: At the MOOC crossroads where the interests of academia and business converge. EMOOCs 2019* (Lecture Notes in Computer Science, Vol. 11475, pp. 37–46). Springer.

Consortium for School Networking (CoSN). (2023). Driving K-12 innovation: 2023 hurdles, accelerators, and recommendations for state education technology and instructional leaders. https://www.cosn.org/tools-and-resour

ces/resource/leading-and-implementing-educational-technology-innovations-hurdles-accelerators-and-tech-enablers-webinar-recording/

Coopersmith, K. (2022). Personal development planning and vertical leadership development in a VUCA World. *The Journal of Values-Based Leadership, 15*(1), 10.

Cormier, D., & Siemens, G. (2010). The open course: Through the open door—open courses as research, learning, and engagement. *EDUCAUSE Review, 45*(4).

Damayanti, F. P., & Mirfani, A. M. (2021, February). An analysis of digital leadership in the pandemic Covid-19 ERA. In *4th international conference on research of educational administration and management (ICREAM 2020)* (pp. 156–159). Atlantis Press.

DeakinCo. (2017). *Soft skills for business success* [online] (pp. 1–36). Deloitte Access Economics. Available from https://www2.deloitte.com/content/dam/Deloitte/au/Documents/Economics/deloitte-au-economics-deakin-soft-skills-business-success-170517.pdf

De Vries, P., & Brall, S. (2008). Microtraining as a support mechanism for informal learning. In *Elearningpapers of Elearningeuropa*. http://www.academia.edu/download/30907447/dcart.pdf

Egloffstein, M., & Ifenthaler, D. (2017). Employee perspectives on MOOCs for workplace learning. *TechTrends, 61*(1), 65–70.

Egloffstein, M. (2018). Massive open online courses in digital workplace learning: Current state and future perspectives. *Digital workplace learning: Bridging formal and informal learning with digital technologies* (pp. 149–166). Springer.

Egloffstein M., Schwerer F. (2019) Participation and achievement in enterprise MOOCs for professional development: Initial findings from the open SAP University. In D. Sampson, J. Spector, D. Ifenthaler, P. Isaías, & S. Sergis (Eds.), *Learning technologies for transforming large scale teaching, learning, and assessment* (pp. 91–103). Springer.

Eşfer, S. (2020). *Instructional design process dynamics of pdMOOCs* [Unpublished doctoral dissertation]. Middle East Technical University. https://open.metu.edu.tr/handle/11511/89632

Esfer, S., & Cagiltay, K. (2018). Creating a MOOC portal for workplace learning. In G. Ifenthaler (Ed.), *Digital workplace learning* (pp. 167–185). Springer. https://doi.org/10.1007/978-3-319-46215-8_10

European Commission, Directorate-General for Education, Youth, Sport and Culture. (2019). *Key competences for lifelong learning*. Publications Office. https://data.europa.eu/doi/10.2766/569540

Fassinger, R. E. (2019). On community, courage, and change—and falling in love. *The Counseling Psychologist, 47*(8), 1136–1142.

Hargreaves, A., & Fink, D. (2018). *Collaborative professionalism: When teaching is a team sport.* Sage.

Haemer, H. D., & Borges-Andrade, J. E. (2018). Learning strategies at work scale: A confirmatory factor analysis. *Psico-USF, 23*(4), 609–620.

Ho, A. D., Chuang, I., Reich, J., Coleman, C. A., Whitehill, J., Northcutt, C. G., Williams, J. J., Hansen, J. D., Lopez, G., & Petersen, R. (2015). HarvardX and MITx: Two years of open online courses Fall 2012–Summer 2014. *SSRN Electronic Journal.* https://doi.org/10.2139/ssrn.2586847

Holley, D., Coulson, K., Buckley, C., & Corradini, E. (2022). Wellbeing in the workplace: Exploring the VUCA approach. *Journal of Learning Development in Higher Education, 25.*

Hoff, D. F., & Smith, D. E. (2020). Leadership and learning agility: A lifelong journey for W. Warner Burke. *The Journal of Applied Behavioral Science, 56*(4), 492–502.

Hickman, L., & Akdere, M. (2018). Effective leadership development in information technology: Building transformational and emergent leaders. *Industrial and Commercial Training, 50*(1), 1–9.

International Society for Technology in Education (ISTE). (2014). *ISTE Standards for educators.* Retrieved February 24, 2018, from https://www.iste.org/standards/for-educators

International Society for Technology in Education (ISTE). (2022). *ISTE Standards for educators.* Retrieved March 14, 2023, from https://www.iste.org/standards/for-educators

Kay, J., Reimann, P., Diebold, E., & Kummerfeld, B. (2013). MOOCs: So many learners, so much potential.... *IEEE Intelligent Systems, 28*(3), 70–77.

Kop, R., & Carroll, F. (2011). Cloud computing and creativity: Learning on a massive open online course. *European Journal of Open, Distance and E-Learning, 14*(2).

Liyanagunawardena, T. R., Adams, A. A., & Williams, S. A. (2013). MOOCs: A systematic study of the published literature 2008–2012. *International Review of Research in Open and Distributed Learning, 14*(3), 202–227.

Moraes, V. V., & Borges-Andrade, J. E. (2015). Individual and contextual variables in municipal officers' workplace learning. *Journal of Workplace Learning, 27*(2), 95–111. https://doi.org/10.1108/JWL-02-2014-0020

National School Boards Association. (2022). Leading the digital leap: Preparing superintendents and school boards for the digital age. https://www.nsba.org/-/media/NSBA/File/Reports/Digital-Leap/digitalleap_final_web.ashx

Nasir, M. K. M. (2020). The influence of social presence on students' satisfaction toward online course. *Open Praxis*, *12*(4), 485–493.

OpenAI. (2023). GPT-3.5. https://openai.com/research/

Popova, N., & Shynkarenko, V. (2016). Personnel development at enterprises with regard to adaptation to the VUCA world. *Економічний часопис-XXI*, *156*(1–2), 88–91.

Patru, M., & Balaji, V. (2016). *Making sense of MOOCs. A guide for policy-makers.* United Nations Educational, Scientific and Cultural Organization (UNESCO). http://unesdoc.unesco.org/images/0024/002451/245122E.pdf

Raja, A. S. (2021). Business research in the VUCA world. *Ushus Journal of Business Management*, *20*(1), v–xvi. https://doi.org/10.12725/ujbm.54.0

Sablina, S., Kapliy, N., Trusevich, A., & Kostikova, S. (2018). How MOOC-takers estimate learning success: Retrospective reflection of perceived benefits. *The International Review of Research in Open and Distributed Learning*, *19*(5). https://doi.org/10.19173/irrodl.v19i5.3768

Sarkar, A. (2016). We live in a VUCA world: The importance of responsible leadership. *Development and Learning in Organizations*, *30*(3), 9–12. https://doi.org/10.1108/DLO-07-2015-0062

Sinha, D., & Sinha, S. (2020). Managing in a VUCA world: Possibilities and pitfalls. *Journal of Technology Management for Growing Economies*, *11*(1), 17–21. https://doi.org/10.15415/jtmge.2020.111003

Small, J., Deacon, A., Walji, S., Jaffer, T., & Jawitz, J. (2019). Building capabilities: Using MOOCs to make transitions in work. *Open Praxis*, *11*(4), 427.

Voogt, J., Knezek, G., Cox, M. J., Knezek, D., & ten Brummelhuis, A. (2017). Under which conditions does ICT have a positive effect on teaching and learning? A call to action. *Journal of Computer Assisted Learning*, *33*(1), 1–5.

Walia, K. (2020). Benefits and challenges of MOOCs: An overview. *Studies in Indian Place Names*, *40*(33), 18–20.

WEF. (2020). *The future of jobs report 2020.* World Economic Forum Report, https://www.weforum.org/reports/the-future-of-jobs-report2020

WEF. (2022). *Strategic intelligence.* World Economic Forum. https://intelligence.weforum.org/topics/a1Gb0000000LPFfEAO/keyissues/a1Gb00000015QJoEAM

WEF. (2023). https://intelligence.weforum.org/topics

Witthaus, G., Inamorato, A., Santos, D., Childs, M., Anne-Christin, Tannhäuser, Conole, G., Bernard, Nkuyubwatsi, & Punie, V. (2016). *Validation of non-formal MOOC-based learning: An analysis of assessment and recognition practices in Europe (OpenCred)*. European Commission. https://doi.org/10.2791/809371

Zawacki-Richter, O., Bozkurt, A., Alturki, U., & Aldraiweesh, A. (2018, February 23). What research says about MOOCs—An explorative content analysis. *The International Review of Research in Open and Distributed Learning, 19*(1). https://doi.org/10.19173/irrodl.v19i1.3356

Zimmer, W. K., McTigue, E. M., & Matsuda, N. (2021). Development and validation of the teachers' digital learning identity survey. *International Journal of Educational Research, 105*, 101717.

Dr. Meltem Özmutlu

The Future Leaders: Trends, Issues and New Solutions

Abstract
Individuals have developed new products and technologies to make their lives easier and to increase the efficiency of all the work they have done. These developments have become subjects that people need to use not only in their daily lives but also at different points in their lives. It is possible to say that education is at the beginning of these issues. The aims, policies and practices of education in developed world countries are undergoing a transformation with these developments. It is observed what the outputs of these transformations are and how they affect them, and steps are taken accordingly. In other words, it is possible to integrate all technological developments and future trends that affect the stages or content of education within institutions. However, in order to make this follow-up, educational leaders must have a command of these developments and be able to use the innovations that come with these developments in their lives. The fact that educational leaders who dominate the fields of artificial intelligence, cloud technologies, big data and other trending topics, demonstrate visionary leadership and create a digital age of learning. The digital culture that leaders will bring to the individuals they work with means an understanding of the progress and development of the corporate culture in this direction. This chapter aims to help educational leaders understand how to use and integrate artificial intelligence, big data and cloud technologies within their organizations. Designs to be made and digital trends to be used to identify individual differences and solve problems in the processes of learning, teaching and management are a subject that education leaders and all educators who are together with them focus on. In the chapter, problems and requirements of today and future trends such as artificial intelligence, big data and cloud technologies and their interactions with each other will be discussed. At the end of this section, educational leaders will learn to be the main character who will recognize the issues that are considered as the future trends of educational technologies and that will affect all steps of education, have knowledge about their use and provide integration into educational environments.

Keywords: Artificial intelligence, big data, cloud computing

The digital age creates quite different changes in human life thanks to the technological developments it contains. All the tools, products and structures developed are based on making people's lives easier and increasing the quality and efficiency of the work done. All technological needs are developed, have a structure in direct proportion with the needs and wishes of individuals in the process, and have a structure that is constantly differentiating, developing and becoming widespread. These observed changes lead to small or large-scale changes and transformations in people's habits and knowledge in society (Korucu & Biçer, 2020). With the impact of the digital age, attention should also be given to the changing role of educational leaders. Educational leaders are now defined not only as teachers' managers but also as leaders who increase students' learning potential by integrating technology into their learning processes. Educational leaders in the digital age should ensure that students take an active role in the learning process by combining their teaching strategies with technology. For example, it is possible to increase students' participation in the learning process and make their learning more enjoyable by using digital tools (Johnson et al., 2016). In line with all these, the leader and leadership are described as the most important elements in the institution. Technological innovations and developments enable educational leaders to create a new vision and acquire new skills regarding the developing structure (Yusof et al., 2019).

Educational leadership is important for understanding the impact of technology development on schools and integrating these developments into instructional strategies and programs. It is an indisputable fact that technology plays an important role in the learning process. However, the use of technology not only supports the learning process of students but can also help educational leaders make all processes in their schools more efficient and effective. Therefore, educational leaders must put technology at the center of their teaching and learning strategies by providing the tools needed to increase students' learning potential (Wang et al., 2011). This section will present explanations and examples of how educational leaders can use the information on topics such as artificial intelligence, big data and cloud computing systems, which are described as current technological developments and trends.

Educational Leaders and Artificial Intelligence (AI)

Although much has changed over the years, artificial intelligence (AI) has remained the same as its original purpose and is characterized as the creation of programs that can grasp what people are thinking and provide as a close approach as possible (Shukla Shubhendu & Vijay, 2013). Artificial intelligence is defined as an algorithm or a computerized system that resembles human characteristics or thinking processes such as generalization, reasoning, decision-making, meaning-making and learning from past experiences (Castelvecchi, 2016). It draws attention as an aspect emphasized in the definitions that these features of artificial intelligence, which is stated to have these human-specific features or processes, are dynamic (Nabiyev & Erümit, 2020). In another definition, it is stated that AI is a branch of computer science that provides a combination of terms that can also be described as sub-branches of AI such as machine learning, algorithm development and natural language processing (Akgün & Green, 2021). Joshi (2020), on the other hand, states that it is artificial intelligence to build structures consisting of human-like machines or systems that do not have extraordinary intelligence from solving the problems encountered. Baker and Smith (2019) provide a comprehensive definition by describing computers that perform cognitive tasks such as learning and problem-solving, which are associated with the human mind, such as AI. This definition is also an indication that artificial intelligence cannot be explained by only one technology. It acts as an umbrella concept describing different technologies and methods such as machine learning, natural language processing, neural networks, data mining and algorithms (Zawacki-Richter et al., 2019).

There is the fact that AI has been defined, as a developing technology and is a part of life by living conditions. The development of AI in countries has a structure that integrates into areas that are considered important in human life such as education, health and communication (Popkova & Sergi, 2020; Vazhayil et al., 2019). AI is described as the harbinger of an unprecedented wave of automation, and it is stated that it causes changes in professions and industries and also plays a role in shaping future jobs (Frank et al., 2019). In addition, it is accepted that AI is a critical technology for overcoming the problems in the global context that will be encountered in the future (Kaplan & Haenlein, 2020). Taking into account all of these, it is mentioned that individuals of all ages should be included in the teaching of artificial intelligence at all educational levels from early childhood to high school for current or future world problems (Sanusi & Olaleye, 2022).

It is stated in the literature that there is a need for more research on the use of AI in education (Zawacki-Richter et al., 2019). However, it is stated that this technology has also increased its capabilities significantly and has reached such a wide potential to be used in all areas of life (Tyson & Sauers, 2021). As in other fields, AI is thought to revolutionize education, such as helping to improve course content, teaching methods and course materials (Yousuf & Wahid, 2021). The developments in AI technology motivate teachers and all other stakeholders in education to shape them in line with their own needs, instead of being just users of advanced technological products (Tunalı, 2018). When talking about the transformation of AI in education, three main headings are mentioned: administrative task automation, intelligent content creation and smart educational systems. With administrative automation, it is mentioned that artificial intelligence will be tasked with evaluating students' assignments and exam results instead of educators' (Tahiru, 2021). Johnson (2019) discusses the use of AI in the admission process of students to schools, which can be described as an evaluation process.

Supporting teachers and supporting students' learning to make their experiences meaningful is possible with the use of AI in education (How & Hung, 2019). The field of artificial intelligence and in line with the opportunities it offers, provides different effects with the different perspectives of the stakeholders in education, as it has the potential to take place in administrative or all different parts of the educational environment (Kuprenko, 2020). While it is stated that the point of collecting, managing, and presenting the data in education by AI is supportive in the education management part, the ways and roles it offers about supporting the learner are stated as its support in the learning and teaching processes (Arslan, 2020). Although computer-aided instructional programs (CAI) are not accepted as true AI content, they are accepted as one of the first systems that show similarity. These systems ask questions to the students and inform the correctness of the answers given by the students and continue the study system (McArthur et al., 2005). In the structure that started with this system, feedback within the traditional education system was integrated into CAI and transformed into smart teaching systems. With this integration, it is considered an important step as personal formative data are presented during teaching (Shukla Shubhendu & Vijay, 2013). Dickson (2017) states that the developed software programs can store the student's pre-test and post-test data, but AI-supported programs not only store them but also enable them to determine which aspects of the curriculum student struggles with by staying longer.

Coccoli et al. (2016) emphasize that AI is currently used for educational purposes and can be used for administrative purposes in education. With its capabilities, AI has the potential to have a major impact on administrative tasks and data analysis. In this case, the use of AI in education is an indication that teachers focus less on evaluating and processing information if they have AI to help (Huang et al., 2019). Coccoli et al. (2016) emphasize that artificial intelligence has become an area where teachers can use or receive support during classroom management. AI is the field that has the greatest impact on students' learning, as well as a support that will create opportunities for teachers to achieve the highest efficiency in doing their jobs (Tyson & Sauers, 2021).

Research highlights that knowing technology is a component of being an effective and good school leader, but there is a lack of research on school technology leadership (McLeod & Richardson, 2011). It requires organizations to experience different types of management to overcome the change brought about by AI (Haung et al., 2019). In line with this reality, educational leaders should be aware of the need to engage deeply in business related to the technological aspects of their schools. The rapid increase in the use of technology in education brings along the need for educational leaders to develop and keep up in these areas. Persichitte (2013) mentions that both research and tools in educational technologies are progressing at a pace that cannot be caught completely, and that leadership should be done by identifying the risks related to this development. The International Society for Technology in Education demonstrates the importance of this situation, especially by setting ISTE standards that specify the knowledge and behaviors that enable educational leaders to empower their teachers and enable students to learn (Tyson & Sauers, 2021). Different types of management are required to overcome the change that AI creates and will create within institutions (Haung et al., 2019). For this reason, unless educational leaders find direction in the new reality, change will not be possible in their institutions. This is only possible when educational leaders are deeply involved in the technological aspects of their institutions (Hamzah et al., 2014). Educational leaders need to be strategic in their approach to adopting AI and applying it in their organizations and be active participants in the adoption and implementation phases. While this means that leaders focus on their learning, as well as communicating with others, it also enables them to establish a successful structure (Tyson & Sauers, 2021).

AI has the potential to improve data literacy in collecting, examining, analyzing and interpreting data for making and reporting decisions in educational environments where educational leaders are (Gummer & Mandinach, 2015). On the other hand, it is known that many educational leaders have insufficient training on how to collect, analyze and interpret data (Lasater et al., 2019; Luo, 2008; Mandinach & Gummer, 2013). The data literacy of educational leaders is described as developed with AI. Individuals need to identify and prioritize how AI works, what opportunities it creates, and know what the limits of AI are. Thanks to the basic artificial intelligence knowledge and examples gained, educational leaders, develop data literacy skills by gaining expert knowledge in data analytics such as up-to-date statistical modeling and analytical methods such as text mining, probabilistic subject modeling, sentiment analysis, network analysis and convolutional network analysis (visual information analysis) (Wang, 2021). Bowers et al. (2019) indicate that in addition to gaining data literacy individually, it is very important for leaders to share responsibilities for adopting and adapting by forming a core team with teachers and other stakeholders to use AI in the data-based decision-making process. In addition, it is mentioned that the stakeholders who are responsible for everything from the education system to the institution building should take place within the institutions, and that education strategies should be developed with artificial intelligence and that it is necessary to work with artificial intelligence services and software providers.

The roles and responsibilities of leaders are of great importance in the creation of AI within the institution, the determination of strategies regarding the created AI, making decisions and supervising its operation in the process. Since the creation of self-managed AI systems, the necessity of using a wide variety of expertise and social skills has come to light, bringing together many decision-makers and process managers, including different thoughts, working styles, different behaviors and experiences, etc. and making the process collaborative. Collective leadership is needed. In addition, it is thought that neuroscience leadership can make valuable contributions to ensuring harmony between AI, leaders, and employees based on neuroscience data (Ülgen & Yavuz, 2021).

Leaders have an active role in ensuring the organization of the use of artificial intelligence within the organization. In this process of use, it also acts as a facilitator when it comes to researching and discussing the effect of artificial intelligence on students and teachers, and the obstacles encountered in the adoption of artificial intelligence are also decisive. According to all leader's determinations, artificial intelligence should develop its capabilities in places of use and guide the changes in the basic organizational culture. The execution of a system in this way and the creation of organizational capacity may mean that the educational leader makes the structure more beneficial, which is likely to spend time and money in her/his institution (Fountaine et al., 2019). Tyson and Saures (2021), on the other hand, state that leaders should be strategic in their approach to adopting and applying AI, and they are role players in creating structures for success in their organizations by focusing on their learning and making connections with others instead of being passive participants.

Educational Leaders and Big Data

In line with the necessity of the age, technological developments are of great importance, and it is possible to store different data such as numerical, textual and graphical that can be used for needs through computers. It is possible to obtain different data in various fields, and there is a large amount of information about students in schools and this information needs to be stored for a long time. Obtaining information that will be considered valuable in line with this information, sometimes in a complex form, and using them by distinguishing and classifying them is defined as data science (Leek, 2013). From space sciences to the field of medicine, advanced technologies are used, and thanks to the machines they buy, information can be collected in a short time thanks to data science and new information can be produced (Lepine & Shara, 2005).

Big data is an idea and innovation that is considered new for individuals engaged in information management (Roy & Singh, 2017). It is a concept that describes a heterogeneous data set or sets with different volumes that are characterized as impossible to process by using database techniques that are characterized as classical (Gahi et al., 2016). Ward and Barker (2013) note that the most common definition of the concept of big data is the form included in the Gartner report and defined by the abbreviation 3V. Regarding the 3V in this definition, Furht and Villanustre (2016) describe it as the three basic features that big data is defined. Different sources in the literature state that three main components characterize the big data phenomenon: variety, velocity and volume as given in Fig. 1 (Golov & Rönnbäck, 2017; Kuiler, 2014; McNeely & Hahm, 2014; Sağıroğlu & Sinanç, 2013).

Volume: It is characterized as millions of devices and applications such as smartphones, social networking applications, sensors and product codes constantly producing large digital data in volume (Oussous et al., 2018). Volume is cited as the main feature of big data problems. The main reason for this is the necessity of innovative tools at the point of data scale, data storage and data access (Cyganek et al., 2016). Cyganek et al. (2016) approached this component of

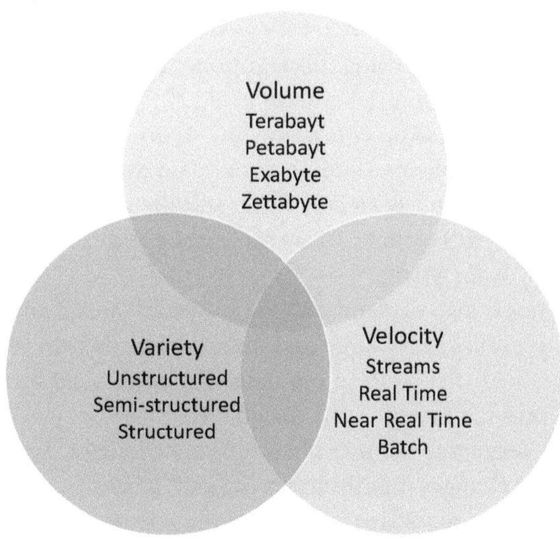

Fig. 1. The Three Vs of Big Data

big data with sampling from the field of health within the scope of their studies. In addition to having the notes of a doctor's patient, there are examination files and images of the patient that should be kept. The area needed is increasing with the increase in the materials that need to be stored. In addition, if the patient's past data is required to be used in the analysis, the data volume will inevitably increase to exabyte sizes. In this case, it creates the necessity of innovative tools at the point of storing and accessing data.

Variety: Big data can produce data in different types and formats thanks to its wide range, and there is no standardized rule or sequence among these mixed data types. In diversity, data is divided into three: structural, semi-structured and unstructured (Gahi et al., 2016; Gandomi & Haider, 2015; Sağıroğlu & Sinanç, 2013). In line with all these, diversity in big data is characterized as structural heterogeneity in the dataset (Aktan, 2016). When it comes to examples, all the data stored in different formats and places such as e-mails, documents stored in image format by scanning, real transaction records in the database, logs, and voice recordings of customer contact centers are valuable for certain institutions. Examining all these different data sets obtained requires the use of different technologies and approaches (Atan, 2016).

Velocity: Data is constantly in motion. The fact that data is constantly in motion has also made data flow analysis an important topic for data scientists (Cyganek et al., 2016). Speed is an important factor not only for big data, but also for all business processes in it, and the speed of production of big data over time is very high and this speed increases as time progresses. There are various examples in the literature regarding this component. Sharing 2.7 billion likes and comments per day on Facebook (Gerhardt et al., 2012) and processing 50 billion messages per day on WhatsApp (Burnham, 2014) are examples of social media. The fact that a jet plane collects 10 terabytes of data through its sensors in every 30 minutes it has flown (Sun & Heller, 2012) and the generation of 20 gigabytes of data by 150 sensors in Formula 1 cars are among the examples outside of social media (George et al., 2014).

Big data analytics is also explained as the use of advanced analytical and parallel techniques to process very large and diverse records with different types of content that came with the birth of big data. The tools used for the realization of this analytics provide benefits and insights that can be qualified as valuable by dealing with a very large volume of complex, structured, semi-structured or structured data that changes rapidly and is difficult to process if traditional databases are used (Gahi et al., 2016). Although the use of big data in the field of education is limited, the tendency to use big data in the decision-making process is constantly increasing (Bughin et al., 2010). This increase in big data in the decision-making phase is due to the efficiency of data-based decision-making. It is thought that this trend will continue to grow in the future and educational institutions will need experts who can process large-scale data and technologies based on big data. Therefore, educational leaders in educational institutions need to be familiar with these new tools and technologies or create departments or business areas to facilitate data-driven decision-making (Kalim, 2021).

Fisher et al. (2020), in their study, explain the benefits of big data in education at three levels: micro level, medium level and macro level. Micro-level associates with keywords clickstream data, medium-level text data and macro-level corporate data. Micro-level big data is automatically collected during interactions between students and related learning environments from platforms such as smart teaching systems, massive online open courses (MOOC), simulations and games. Therefore, micro-level big data is characterized as the interaction data between the actions taken by the learner, and which can capture the individual data of the learners. Intermediate big data includes learner writings in which students' writing activities are systematically collected, from their in-class assignments to their participation in online discussion forums and social media interactions. These data provide raw data, particularly on students' emotional states and the progress of their cognitive and social abilities. Demographic information of students, admission information, service data of institution campuses, class schedules and course enrollment data are included in macro-level big data. Therefore, it is possible to say that the data collected at the institutional level constitutes the big data at the macro level. The collection interval of such data is within the scope of the annual time frame and is updated once or twice at the beginning of the period.

With the increase in the use of big data in recent years, this data set open new doors to how we monitor, understand and evaluate processes in educational contexts and play a role in improving the effectiveness of education (Fisher et al., 2020). It is stated that big data is generally data on the administrative part or learning process in the context of education, and each of these data has offered its promises for the scope of educational research (National Academy of Education, 2017). Fisher et al. (2020) emphasizes that the emergence of big data in education is associated with two main trends in the digital age. The first of these was the use of student information systems to store and organize students' profile information and academic records. These collected data cover decades of student data within the institution and the information system created allows for the management and analysis of the data. The second is the detection of learning behaviors that are difficult to record in face-to-face training and courses by learning management systems (LMS). The LMSs allow instructors to share materials with students, as well as view student assignments and communicate with students. In this system, thousands of data of each student are obtained. Therefore, the use of big data in education helps not only student information systems and LMSs, but also diversifies innovations in digital learning environments, enriches new pedagogical possibilities, and collects the digital footprints of the students in these areas.

Thanks to big data, educational leaders can make timely and effective decisions about improving their corporate culture and learning processes for the learner (Kalim, 2021). According to studies, it is stated that big data has the potential to completely change the education and educational research process in the future. Research has been conducted with charter schools in the United States, and it has been determined that the use of data to guide teaching is one of the five policies associated with significant academic impacts (Picciano, 2012). The concept of big data is considered to have the potential to bring about changes in the means of interaction between students and teachers in the education process and to revolutionize the way that the individual needs of learners are met with efficient processes. In this case, it brings with the necessity of educational leaders to use big data as a valuable information system in education-related decision-making processes and to provide efficiency and effectiveness in decision-making processes (Logica & Magdalena, 2015). In addition, it is big data that supports the classical education system by analyzing what teachers know and what techniques are effective for each student. With this opportunity, teachers can improve the learning process by changing their techniques and methods in the classroom (Kalim, 2021).

Big data is one of the keys to unlocking learning for educational leaders. To effectively implement data-based decisions in the educational environment, educational leaders need to be familiar with the new tools and technologies developed. Technology-based learning platforms are big data providers. The data coming from these sources and the analytics and analyses to be made by the educational leaders can improve the skills of the educational leaders to be more effective and contribute to their being more productive (Falakmasir & Habibi, 2010). There are many advantages of the big data subject that educational leaders will use or should use in the decision-making process. These advantages are divided into efficiency in management work, support in planning and policy, and time savings. Decision-making for the management is one of the main veins of the process, and the decisions made are run efficiently with the proper functioning of the educational institution. Thanks to big data, educational leaders provide timely information and administrative decisions are made with efficiency. Big data provides analysis of the past where it will be used, and also makes predictions for the future. All the information that the educational leaders will receive at this point will be supportive and developers in the planning and policymaking of any educational organization. The advantage of timesaving is an important factor for the success of the institution, choosing the right alternatives or ways in a world where institutions and individuals compete. Big data, on the other hand, assists educational leaders in this accelerating progress in various ways and brings efficiency and effectiveness to the operations carried out within the educational institution (Logica & Magdalena, 2015).

Decision-making is characterized as a complex issue when it is both individual and group level. Decision-making within education is a dynamic process in which educational leaders must make different types of decisions at different times. Therefore, while this dynamic process sometimes includes big or complex sometimes small and simple decisions, it is sometimes involved in decision-making processes that are sometimes unscheduled, while decisions are sometimes made programmatically about its planning. The purpose of all these decisions that the training leader should take is to increase the efficiency and productivity of the institution or organization by choosing the best one in the conditions (Logica & Magdalena, 2015).

Educational Leaders and Cloud Computing

Cloud computing technologies are defined as a model that can be provided quickly with service-provider interaction and provide easy access to computing resources over the network, without requiring high effort from management (Mell & Grance, 2011). Another definition states that cloud computing is a technology where personal data is stored, you can run applications on it despite the need for high hardware, and you can access web-based applications without the concept of time (Kozan et al., 2014). Although cloud technology has a non-standard structure, there are types of services offered by cloud computing providers. These are (Mell & Grance, 2011):

Software as a Service (SaaS): With this service, users can access all the applications they need over the web. It is also described as the most used type in education. Communication and office applications such as e-mail, and spreadsheets are examples of this type.
Platform as a Service (SaaS): It provides a service based on the ability to operate systems, databases, middleware and web servers to use remote clients. Therefore, there are service offerings such as testing, sharing and repairing applications over the internet.
Infrastructure as a Service (SaaS): This service works for payment logic as much as it is used. More memory, bandwidth and data storage, etc. It is the type of service used for transactions. In this type of service, a kind of virtual hardware is offered to users.

Cloud computing is a technology that is used in education and can sometimes be described as a big data technology dimension. Cloud computing, which is described as collaborative technology, allows the development of educational services such as providing cheap access to online content to students in need (Bughin et al., 2010). Cloud computing provides educational leaders with many opportunities to support their teachers and students. Thanks to cloud computing, students and teachers can easily interact with each other and increase the efficiency of cooperation. This technology also offers advantages in terms of ensuring data security, easy backup and accessibility. Educational leaders can support their teachers in their institutions on how to use digital tools through cloud computing and create virtual learning environments to enable students to learn outside the classroom (Lee et al., 2008; Sung et al., 2016).

Cloud computing technologies are becoming more and more important in education, thanks to their ability to provide scalable, cost-effective and accessible solutions for different educational applications. Thanks to these features, it offers educational leaders the opportunity to use and transform the methods of providing various technology resources to their schools and students. According to Chang et al. (2018), educational leaders can benefit from using cloud computing through reduced costs, improved scalability, improved accessibility, and increased flexibility. In addition, by making use of cloud computing, the leader can provide his students with access to a wide variety of educational tools and resources from any device and any place with an internet connection. One of the important benefits of this technology is to provide students with the ability to provide personalized learning experiences. With this technology, it is possible to create customized learning environments that can be adapted to the individual needs and preferences of the students, with the leadership of the leaders within the institution. The environment to be created is integrated with the integration of cloud-based learning management systems and different digital tools that provide personalized assessments, teaching materials and feedback. In these environments, it supports collaborative learning by allowing students to work together through projects and assignments, regardless of their physical location. In addition, it encourages teamwork and the development of problem-solving skills, which are important for success. In line with all these, it is thought that the inclusion of cloud computing technologies in education will contribute greatly to the education leaders' creation of a more dynamic, interesting, and effective learning experience for their students (Gikas & Grant, 2013).

Conclusion

Today's educational leaders need to try to fill the gaps in their technical knowledge and skills to lead digital developments in their institutions and to provide guidance and direction to disseminate these applications in their institutions' learning environments (Aksal, 2015). As all technologies become more important in education systems, the role of educational leaders in assuming the role of technological leaders and supporting teachers to integrate technology effectively into the teaching process is emerging. All these roles mean that leaders shape their impact and response to technological changes (Richardson et al., 2015). Leading in this age of continuous technological developments, which is characterized as the age of digital transformation, brings with it the necessity of individuals in the leadership position to be both human-oriented and technically careful (Diamante & London, 2002).

To provide students with a better future, it is crucial for educational leaders to keep up with technological advancements and innovations. Educational leaders need to be innovative. This necessity of being innovative brings with it the ability to take initiative for educational leaders. In the initiatives taken, it allows him/her to turn the situations around him into an opportunity and to improve himself/herself. Creativity is another feature that leaders should have, and it means reinterpreting the situations encountered from a different perspective. In order not to miss the opportunities in the encountered situations, it is necessary to take quick decisions and take immediate action for both leaders and the people under their control (Bayram, 2013).

Leaders must adopt strategic ways when implementing AI in their institutions when it comes to the usage of this technology in education. They should be aware that it is not appropriate to actively participate in conversations about AI and be passive participants in the adoption and subsequent implementation process. All of these require them to focus on their learning and at the same time communicate with others. If leaders use AI technology in their organizations, it can not only improve student learning but also reduce teachers' workload. Teachers can accelerate the process by using AI technologies instead of preparing customized training programs to meet the personal learning needs of students. With a chatbot to be used, certain questions of the students can be answered in a shorter time, and it may be possible to prepare a learning plan specific to the student by easily following the learning styles of the students with machine learning (Nikou & Economides, 2019).

With the development of big data, educational leaders will not have to face traditional difficulties in decision-making. Because the use of big data techniques helps educational leaders make accurate and timely decisions. Therefore, big data has the potential to shape education systems in a modern and dynamic way. In addition, educational leaders need to improve the teaching and learning process in their institutions while using and making use of this technology (Kalim, 2021). Cloud computing is a technology that is used in education and can sometimes be described as a big data technology dimension. Cloud computing, which is described as collaborative technology, allows the development of educational services such as providing cheap access to online content to students in need (Bughin et al., 2010). Therefore, cloud computing, like artificial intelligence and big data, has a structure that educational leaders should use actively in their institutions, which can both support the digitalization of teachers and support students' learning outside the classroom (Lee et al., 2008; Sung et al., 2016).

Educational leaders must be at peace with developing and changing technological orders and tools and must take on the role of technological leader for everyone in the institution. The main reason for this is knowing the fact that changes can occur in institutions only with the use, integration, and guidance of leaders. It is important to decide how and where technologies such as artificial intelligence, big data, and cloud computing should be used well and to use them at points that will provide efficiency in learning environments. It should be known that providing the right integration and guidance as the control leaders at the point of use and learning of all technological developments will create significant changes in the future.

References

Aksal, F. A. (2015). Are headmasters digital leaders in school culture? *Eğitim ve Bilim*, *40*(182). https://doi.org/10.15390/EB.2015.4534

Aktan, E. (2018). Büyük veri: Uygulama alanları, analitiği ve güvenlik boyutu. *Bilgi Yönetimi*, *1*(1), 1–22. https://doi.org/10.33721/by.403010

Arslan, K. (2020). Eğitimde yapay zekâ ve uygulamaları. *Batı Anadolu Eğitim Bilimleri Dergisi*, *11*(1), 71–88.

Atan, S. (2016). Veri, büyük veri ve işletmecilik. *Balıkesir Üniversitesi Sosyal Bilimler Enstitüsü Dergisi*, *19*(35), 137–154. https://doi.org/10.31795/baunsobed.645312

Bayram, Ş. (2013). *Liderlik kavramı ve liderlik türlerinin inovasyon üzerindeki etkileri* [Yayımlanmamış yüksek lisans tezi]. Gebze Yüksek Teknoloji Enstitüsü, Gebze.

Bowers, A. J., Bang, A., Pan, Y., & Graves, K. E. (2019). Education leadership data analytics (ELDA): A White Paper Report on the 2018 ELDA Summit. *Online Submission*. https://doi.org/10.7916/d8-31a0-pt97

Bughin, J., Chui, M., & Manyika, J. (2010). Clouds, big data, and smart assets: Ten tech enabled business trends to watch. *McKinsey Quarterly*, *56*(1), 75–86.

Burnham, K. (2014, 21 February). *Facebook's whatsApp buy: 10 staggering stats.* Retrieved February 2, from https://www.informationweek.com/social/facebook-s-whatsapp-buy10-staggering-stats

Castelvecchi, D. (2016). Can we open the black box of AI? *Nature News*, *538*(7623), 20. https://doi.org/10.1038/538020a

Chang, C. Y., Lai, C. L., & Hwang, G. J. (2018). Trends and research issues of mobile learning studies in nursing education: A review of academic publications from 1971 to 2016. *Computers & Education*, *116*, 28–48. https://doi.org/10.1016/j.compedu.2017.09.001

Coccoli, M., Maresca, P., & Stanganelli, L. (2016). Cognitive computing in education. *Journal of E-learning and Knowledge Society*, *12*(2), 55–69. https://doi.org/10.20368/1971-8829/1185

Cresswell, K., Callaghan, M., Khan, S., Sheikh, Z., Mozaffar, H., & Sheikh, A. (2020). Investigating the use of data-driven artificial intelligence in computerised decision support systems for health and social care: A systematic review. *Health Informatics Journal*, *26*(3), 2138–2147. https://doi.org/10.1177/1460458219900452

Cyganek, B., Graña, M., Krawczyk, B., Kasprzak, A., Porwik, P., Walkowiak, K., & Woźniak, M. (2016). A survey of big data issues in electronic health record analysis. *Applied Artificial Intelligence*, *30*(6), 497–520. https://doi.org/10.1080/08839514.2016.1193714

Diamante, T., & London, M. (2002). Expansive leadership in the age of digital technology. *Journal of Management Development*, *21*(6), 404–416. https://doi.org/10.1108/02621710210430597

Dickson, B. (2017). How artificial intelligence is shaping the future of education. *PC Magazine*. Retrieved February 2, from https://www.pcmag.com/news/how-artificial-intelligence-is-shaping-the-future-of-education

Falakmasir, M. H., & Habibi, J. (2010, June). Using educational data mining methods to study the impact of virtual classroom in e-Learning. In *Educational data mining (EDM) conference* (pp. 241–248).

Fischer, C., Pardos, Z. A., Baker, R. S., Williams, J. J., Smyth, P., Yu, R., Slater, S., Baker, R., & Warschauer, M. (2020). Mining big data in education: Affordances and challenges. *Review of Research in Education*, *44*(1), 130–160. https://doi.org/10.3102/0091732X20903304

Fountaine, T., McCarthy, B., & Saleh, T. (2019). Building the AI-powered organization. *Harvard Business Review*, *97*(4), 62–73.

Frank, M. R., Autor, D., Bessen, J. E., Brynjolfsson, E., Cebrian, M., Deming, D. J., Feldman, M., Groh, M., Lobo, J., Moro, E., Youn, H., & Rahwan, I. (2019). Toward understanding the impact of artificial intelligence on labor. *Proceedings of the National Academy of Sciences*, *116*(14), 6531–6539. https://doi.org/10.1073/pnas.1900949116

Furht, B., & Villanustre, F. (2016). *Big data technologies and applications*. Springer. https://doi.org/10.1007/978-3-319-44550-2

Gahi, Y., Guennoun, M., & Mouftah, H. T. (2016, June). Big data analytics: Security and privacy challenges. In *2016 IEEE Symposium on Computers and Communication (ISCC)* (pp. 952–957). IEEE. https://doi.org/10.1109/ISCC.2016.7543859

Gerhardt, B., Griffin, K., & Klemann, R. (2012). *Unlocking value in the fragmented world of big data analytics*. Cisco Internet Business Solutions Group. https://www.cisco.com/c/dam/en_us/about/ac79/docs/sp/Information-Infomediaries.pdf

George, G., Haas, M. R., & Pentland, A. (2014). From the editor's big data and management. *Academy of Management Journal*, *57*(2), 321–326. https://doi.org/10.5465/amj.2014.4002

Gikas, J., & Grant, M. M. (2013). Mobile computing devices in higher education: Student perspectives on learning with cellphones, smartphones & social media. *The Internet and Higher Education*, *19*, 18–26. https://doi.org/10.1016/j.iheduc.2013.06.002

Golov, N., & Rönnbäck, L. (2017). Big data normalization for massively parallel processing databases. *Computer Standards & Interfaces*, *54*, 86–93. https://doi.org/10.1016/j.csi.2017.01.009

Gummer, E. S., & Mandinach, E. B. (2015). Building a conceptual framework for data literacy. *Teachers College Record*, *117*(4), 1–22. https://doi.org/10.1177/016146811511700401

Hamzah, M. I. M., Juraime, F., Hamid, A. H. A., Nordin, N., & Attan, N. (2014). Technology leadership and its relationship with school-Malaysia standard of education quality (School-MSEQ). *International Education Studies*, *7*(13), 278–285. https://doi.org/10.5539/ies.v7n13p278

How, M. L., & Hung, W. L. D. (2019). Educing AI-thinking in science, technology, engineering, arts, and mathematics (STEAM) education. *Education Sciences*, *9*(3), 184. https://doi.org/10.3390/educsci9030184

Huang, M. H., Rust, R., & Maksimovic, V. (2019). The feeling economy: Managing in the next generation of artificial intelligence (AI). *California Management Review, 61*(4), 43–65. https://doi.org/10.1177/0008125619863436

Johnson, A. (2019). 5 ways AI is changing the Education Industry. *ELearning Industry, 15*. Retrieved February 2, from https://elearningindustry.com/ai-is-changing-the-education-industry-5-ways

Johnson, L., Becker, S. A., Cummins, M., Estrada, V., Freeman, A., & Hall, C. (2016). *NMC horizon report: 2016 higher education edition* (pp. 1–50). The New Media Consortium.

Kalim, U. (2021). The growing role of big data in education and its implications for educational leadership. *International Journal of Research and Innovation in Social Science, 5*(1), 257–262. https://doi.org/10.47772/ijriss.2021.5111

Kaplan, A., & Haenlein, M. (2020). Rulers of the world, unite! The challenges and opportunities of artificial intelligence. *Business Horizons, 63*(1), 37–50. https://doi.org/10.1016/j.bushor.2019.09.003

Korucu, A. T., & Bicer, H. (2020). Eğitimde yapay zekanın rolleri ve eğitsel yapay zekâ uygulamaları. V. Nabiyev & A. K. Erumit (Eds.), *Eğitimde yapay zeka kuramdan uygulamaya*. Pegem Akademi.

Kozan, M., Bozkaplan, M. F., & Özek, M. B. (2014). Eğitimde bulut bilişim uygulamaları. *Akademik Bilişim Konferansı, 5*(7).

Kuiler, E. W. (2014). From big data to knowledge: An ontological approach to big data analytics. *Review of Policy Research, 31*(4), 311–318. https://doi.org/10.1111/ropr.12077

Kuprenko, V. (2020). Artificial intelligence in education: Benefits, challenges, and use cases. Retrieved November 26, 2022, from https://pub.towardai.net/artificial-intelligence-in-education-benefits-challenges-and-use-cases-db52d8921f7a

Lasater, K., Albiladi, W. S., Davis, W. S., & Bengtson, E. (2020). The data culture continuum: An examination of school data cultures. *Educational Administration Quarterly, 56*(4), 533–569. https://doi.org/10.1177/0013161X19873034

Lee, M. J., McLoughlin, C., & Chan, A. (2008). Talk the talk: Learner-generated podcasts as catalysts for knowledge creation. *British Journal of Educational Technology, 39*(3), 501–521. https://doi.org/10.1111/j.1467-8535.2007.00746.x

Leek, J. (2013). *Six types of analyses every data science should know* [blog post]. Retrieved February 11, 2022, from http://www.statlit.org/pdf/2013-Smith-Data-Analytics-Six-Types-P1.pdf.

Lépine, S., & Shara, M. M. (2005). A catalog of northern stars with annual proper motions larger than 015 (LSPM-NORTH Catalog). *The Astronomical Journal, 129*(3), 1483. https://doi.org/10.1086/427854

Logica, B., & Magdalena, R. (2015). Using big data in the academic environment. *Procedia Economics and Finance, 33*, 277–286. https://doi.org/10.1016/S2212-5671(15)01712-8

Luo, M. (2008). Structural equation modeling for high school principals' data-driven decision making: An analysis of information use environments. *Educational Administration Quarterly, 44*(5), 603–634. https://doi.org/10.1177/0013161X08321506

Mandinach, E. B., & Gummer, E. S. (2013). A systemic view of implementing data literacy in educator preparation. *Educational Researcher, 42*(1), 30–37. https://doi.org/10.3102/0013189X12459803

McArthur, D., Lewis, M., & Bishary, M. (2005). The roles of artificial intelligence in education: Current progress and future prospects. *Journal of Educational Technology, 1*(4), 42–80.

McLeod, S., & Richardson, J. W. (2011). The dearth of technology leadership coverage. *Journal of school leadership, 21*(2), 216–240. https://doi.org/10.1177/105268461102100204

McNeely, C. L., & Hahm, J. O. (2014). The big (data) bang: Policy, prospects, and challenges. *Review of Policy Research, 31*(4), 304–310. https://doi.org/10.1111/ropr.12082

Mell, P., & Grance, T. (2011). The NIST definition of cloud computing. *National Institute of Standards and Technology*. Retrieved February 6, from https://nvlpubs.nist.gov/nistpubs/legacy/sp/nistspecialpublication800-145.pdf

National Academy of Education. (2017). *Big data in education: Balancing the benefits of educational research and student privacy. A workshop summary.* ERIC Clearinghouse.

Nikou, S. A., & Economides, A. A. (2019). Mobile-based assessment: Integrating machine learning and chatbots into online assessment. *Computers in Human Behavior, 93*, 142–151. https://doi.org/10.1016/j.chb.2018.12.039

Oussous, A., Benjelloun, F. Z., Lahcen, A. A., & Belfkih, S. (2018). Big data technologies: A survey. *Journal of King Saud University-Computer and Information Sciences, 30*(4), 431–448. https://doi.org/10.1016/j.jksuci.2017.06.001

Persichitte, K. A. (2013). Leadership for educational technology contexts in tumultuous higher education seas. *TechTrends, 57*(5), 14–17. https://doi.org/10.1007/s11528-013-0686-5

Picciano, A. G. (2012). The evolution of big data and learning analytics in American higher education. *Journal of Asynchronous Learning Networks, 16*(3), 9–20. https://doi.org/10.24059/olj.v16i3.267

Popkova, E. G., & Sergi, B. S. (2019). *The 21st century from the positions of modern science: Intellectual, digital and innovative aspects*. Springer Nature.

Richardson, J. W., Bathon, J., Flora, K. L., & Lewis, W. D. (2012). NETS• A scholarship: A review of published literature. *Journal of Research on Technology in Education*, *45*(2), 131–151. https://doi.org/10.1080/15391523.2012.10782600

Roy, S., & Singh, S. N. (2017, January). Emerging trends in applications of big data in educational data mining and learning analytics. In *2017 7th international conference on cloud computing, data science & engineering—Confluence* (pp. 193–198). IEEE. https://doi.org/10.1109/CONFLUENCE.2017.7943148

Sagiroglu, S., & Sinanc, D. (2013, May). Big data: A review. In *2013 international conference on collaboration technologies and systems (CTS)* (pp. 42–47). https://doi.org/10.1109/CTS.2013.6567202

Sanusi, I. T., & Oyelere, S. S. (2020, October). Pedagogies of machine learning in K-12 context. In *2020 IEEE frontiers in education conference (FIE)* (pp. 1–8). https://doi.org/10.1109/FIE44824.2020.9274129

Shukla Shubhendu, S., & Vijay, J. (2013). Applicability of artificial intelligence in different fields of life. *International Journal of Scientific Engineering and Research*, *1*(1), 28–35.

Sun, H., & Heller, P. (2012). Oracle information architecture: An architect's guide to big data. *Oracle, Redwood Shores.*

Sung, Y. T., Chang, K. E., & Liu, T. C. (2016). Integrating mobile devices with teaching and learning affect student's learning performance: A meta-analysis and research synthesis. *Computers & Education*, *94*, 252–275. https://doi.org/10.1016/j.compedu.2015.11.008

Tahiru, F. (2021). AI in education: A systematic literature review. *Journal of Cases on Information Technology (JCIT)*, *23*(1), 1–20. https://doi.org/10.4018/JCIT.2021010101

Tunalı, S. (2018). Eğitimde Yapay Zekâ Çalıştayı Sonuç Raporu. Retrieved December 26, 2022, from https://tasam.org/Files/Icerik/File/EYC_Sonuc_TR_pdf_695272bb-076c4edf-a552-adc784201690.pdf

Tyson, M. M., & Sauers, N. J. (2021). School leaders' adoption and implementation of artificial intelligence. *Journal of Educational Administration*, *59*(3), 271–285. https://doi.org/10.1108/jea-10-2020-0221

Ülge, B., & Yavuz, N. (2021). Yapay zeka ile liderliğin dönüşümü. In N. Öykü İyigün & Mustafa K. Yılmaz (Eds.), *Yapay zeka güncel yaklaşımlar ve uygulamalar*. Beta Kitap.

Vazhayil, A., Shetty, R., Bhavani, R. R., & Akshay, N. (2019, December). Focusing on teacher education to introduce AI in schools: Perspectives and illustrative findings. In *2019 IEEE tenth international conference on technology for education (T4E)* (pp. 71–77). https://doi.org/10.1109/T4E.2019.00021

Wang, Q., Chen, L., & Liang, Y. (2011). The effects of technology on learning: A review of recent research. *Journal of Educational Technology Development and Exchange*, *4*(1), 1–14.

Wang, Y. (2021). When artificial intelligence meets educational leaders' data-informed decision-making: A cautionary tale. *Studies in Educational Evaluation*, *69*, 100872. https://doi.org/10.1016/j.stueduc.2020.100872

Yablonsky, S. (2019). Multidimensional data-driven artificial intelligence innovation. *Technology Innovation Management Review*, *9*(12), 16–28.

Yousuf, M., & Wahid, A. (2021, November). The role of artificial intelligence in education: Current trends and future prospects. In *2021 international conference on information science and communications technologies (ICISCT)* (pp. 1–7). IEEE.

Yusof, M. R., Yaakob, M. F. M., & Ibrahim, M. Y. (2019). Digital leadership among school leaders in Malaysia. *International Journal of Innovative Technology and Exploring Engineering*, *8*(9).

Zawacki-Richter, O., Marín, V. I., Bond, M., & Gouverneur, F. (2019). Systematic review of research on artificial intelligence applications in higher education–where are the educators? *International Journal of Educational Technology in Higher Education*, *16*(1), 1–27. https://doi.org/10.1186/s41239-019-0171-0

List of Figures

List of Tables

About the Editor

Sabiha DULAY, PhD, is an assistant professor at the Department of Educational Sciences, Bahçeşehir University, Istanbul, Turkey. She completed her undergraduate degree (2007–2011) in English Language Teaching at Dokuz Eylül University, Izmir, Turkey. Then she received her MA (2011–2013), and PhD (2013–2017) in Educational Administration, Inspection, Planning and Economics at Eskisehir Osmangazi University, Eskisehir, Turkey. Dr. Dulay's main areas of interests are leadership, organizational behavior, educational administration and effective classroom management strategies. So far, she has researched about effective classroom management strategies, leadership styles and organizational behavior. Dr. Dulay has various scientific studies, book chapters, and national/international presentations about leadership, organizational commitment, organizational climate, teacher typologies in classroom management, school principals as socials actors, and educational policies. In addition, she has organized seminars and workshops about conflict and crisis management, effective teachers, classroom management strategies and she has presented her studies in different conferences. She also has been teaching courses in the field of educational sciences at graduate and undergraduate level.

About the Authors

Ceyda AYSUNA TÜRKYILMAZ, born in 1982, graduated from Istanbul University Business Administration Program lectured in English. She graduated from Production Management and Marketing Master Program of Marmara University and started as a research assistant in the same department. She completed her PhD program in Production Management and Marketing department and finished her dissertation in 2011. Ceyda Aysuna Türkyılmaz works as an associate professor of Marketing in Marketing Department at Marmara University Business Administration Faculty. She studies on consumer behavior, marketing, distribution channels and brand management. She has several national and international papers published in journals and presented in conferences along with several national and international books and book chapters published.

Hatice ÇAKIR is a research assistant at Gazi University Graduate School of Educational Science. She graduated from Giresun University Preschool Education Undergraduate Program in 2017 and Ankara University Preschool Education Postgraduate Program in 2022. She is currently a PhD student in preschool education at Gazi University. Her main areas of interest are giftedness in early childhood, media, digital applications and inclusive education. She continues to publish nationally and internationally studies on these subjects.

Berkan ÇELİK, PhD, works as a research assistant Dr. in Van Yuzuncu Yil University (Van YYU) Faculty of Education, Department of Computer Education and Instructional Technology. He completed his undergraduate education at the Middle East Technical University (METU), Faculty of Education, Department of Computer and Instructional Technology in 2011 as the top-ranking student within the department. In 2014, he received his master's degree with the thesis of "Pre-service Teachers' Opinions, Self-efficacy Beliefs and State Anxiety in relation to Educational Podcasting," and he completed his doctorate degree in 2020 with the dissertation of "An Examination of Presage, Process and Product Dimensions in Massive Open Online Courses" from METU Department of Computer and Instructional Technology. He continues to work on online learning, Massive Open Online Courses, gamification and teacher training.

Sezin EŞFER, PhD, is an assistant professor at the Department of Educational Sciences, Bahçeşehir University, Istanbul, Turkey. She earned her BS degree from the Computer Education and Instructional Technology program at Uludag University, Bursa, in 2006 and MS degree in the same department at Marmara

University, İstanbul, in 2010. Then she received PhD at Middle East Technical University, Ankara, Turkey, in 2020. She is coordinator of the Bilgeİş pdMOOC (MOOCs for professional development) portal which were developed specifically for the professional development of employee and employers since 2015. The motivation behind this terminology is instigating continual, free, voluntary, and accessible learning environments for professional development. In light of lifelong learning, workplace learning, informal learning, and open learning, it is a necessity to integrate instructional elements, approaches, and strategies to design a more need-oriented form of MOOCs. The details of pdMOOCs can be found at Eşfer and Cagiltay's (2018) and Cagiltay, Eşfer, and Çelik's (2020) studies. Generally, her research focuses on human-computer interaction, technology enhanced learning professional development, lifelong learning, teacher education, e-learning, and human-computer interaction, MOOCs. So far, Dr. Eşfer has several experiences and projects about eye tracking usage for effective classroom management and digital competencies of teachers.

B. Çağla GARİPAĞAOĞLU, PhD, is currently working as an Associate Professor at the Faculty of Educational Sciences, Educational Leadership and Administration Program, Bahçeşehir University (BAU), Istanbul, Turkey. She has more than 15 years of experience in teaching and supervision of students, research, consultancy work, workshops, seminars and advisory services within and outside the university. In addition, she has four years of working experience as a school principal at Bahçeşehir College, Umraniye campus which is a home to over 1000 students in Istanbul, Turkey. She completed her undergraduate degree (1996–2001) in Teaching Mathematics department at Boğaziçi (Bosphorous) University (formerly known as Robert College), Istanbul, Turkey. Then she received her MA (2004–2005) in Educational Sciences at The University of Sydney, Sydney, Australia and PhD (2006–2012) in Business Administration at Yeditepe University, Istanbul, Turkey. So far, she has researched about management strategies, leadership styles, and organizational behavior. Dr. Garipaagaoglu has various scientific studies, book chapters, and national/international presentations about organizational leadership & behavior ad specifically interested in conflict management organizational culture and post-modern/quantum leadership.

Hatice Necla KELEŞ, PhD, is a professor at the Department of Educational Sciences, Bahçeşehir University, Istanbul, Turkey. She completed her undergraduate degree (1992) in Business Administration at Selçuk University, Konya, Turkey. Then she received her MBA (1996), and PhD (2006) in Management and Organization at Selçuk University, Konya, Turkey. So far, she has researched about management strategies, human resource management, team management, leadership styles and organizational behavior. Dr. Keleş has various scientific studies, book chapters, and national/international presentations about leadership, organizational commitment, organizational intelligence, new skills and talents in human resource management, generational differences, positive psychological capital, and career development. In addition, she has organized seminars and workshops about effective team management, leadership and coaching.

Özge ÖZTEKİN BAYIR is a faculty member in the Department of Educational Sciences at Eskisehir Osmangazi University, Eskisehir, Turkey. She was graduated from Department of English Language Teaching in Dokuz Eylül University in 2011. She received her M.Sc. (2013) and PhD (2018) in Educational Administration, Inspection, Planning and Economics from Eskisehir Osmangazi University, Department of Educational Sciences. She has offered several undergraduate courses such as Introduction To Education, Classroom Management, Turkish Educational System and School Administration, Comparative Education, Extra Curriculum Activities, Character and Value Education and Teaching Practicum in Faculty of Education. Dr. Öztekin Bayır's main areas of interests are leadership, organizational behavior and educational administration. She conducts research on leadership styles, school culture, organizational commitment, job satisfaction in connection with her areas of interests. In this context, she has national/international book chapters, articles and presentations in different countries about these topics. She is still working at Eskisehir Osmangazi University.

Meltem ÖZMUTLU is a lecturer at Bahçeşehir University, Faculty of Educational Sciences, Department of Computer Education and Instructional Technologies. She completed her undergraduate education in Bahçeşehir University Computer and Instructional Technologies Education between 2011–2016 and in Software Engineering as a double major. Then in 2018, she completed the Game Design Master's Program at Bahçeşehir University, Social Sciences Institute. She has been doing her PhD in Educational Technologies at Bahçeşehir University since 2018 and is at the thesis stage. The researcher, who has publications on computational thinking, data mining and technology integration and school adaptation process, works on artificial intelligence, artificial intelligence literacy, data mining, educational game design, digital game design, digital literacy, distance education and coding education. In her latest works, she focuses on studies to raise awareness of artificial intelligence literacy. In addition to these, the researcher also gave game design and coding trainings for children. She is also involved in projects in the field of educational technologies.

Aybiçe TOSUN received her BA (2007—Theology) and MA (2010—Religious Education) from Ankara University (Turkey). She holds a PhD degree on Religious Education from Ankara University. During her PhD education, she was a visiting researcher at University of New Brunswick, Canada between 2014 and 2015. She is working as an assistant professor at Eskisehir Osmangazi University, Faculty of Theology, Department of Religious Education since 2015. Her main areas of interest are worldview education, multicultural education and creative drama. She improved her experience in multicultural education environments by attending Vienna Intercultural International Summer University (VICISU) in Vienna in 2012 and Christan-Islamic Spring Academy in Tutzing in 2014. She is a member of international religious education societies ISREV (International Seminar on Religious Education and Values) and REA (Religious Education Association).

www.ingramcontent.com/pod-product-compliance
Ingram Content Group UK Ltd.
Pitfield, Milton Keynes, MK11 3LW, UK
UKHW041825031026
15587UKWH00002B/574

9 783631 895962